ESSENTIAL BIOCHEMISTRY, ENDOCRINOLOGY AND NUTRITION

Christine Kingsley.
September 1945.

GW00707710

A guide to important principles
for nurses and allied professions

Essential Biochemistry, Endocrinology and Nutrition

Professor D. F. Horrobin

MTP

MEDICAL AND TECHNICAL PUBLISHING CO LTD
1971

Published by

MTP

Medical and Technical Publishing Co Ltd
Chiltern House, Oxford Road
Aylesbury, Bucks

Copyright © 1971; D. F. Horrobin

SBN 852 0001 38

First published 1971

Books in the 'Essential knowledge'
series for nurses:

Essential Anatomy
Essential Medicine
Essential Physics, Chemistry and Biology
**Essential Biochemistry, Endocrinology
and Nutrition**
Essential Diagnostic Tests
Essential Physiology

PRINTED IN GREAT BRITAIN BY
ALDEN & MOWBRAY LIMITED
AT THE ALDEN PRESS, OXFORD

Contents

THIS SERIES REPRESENTS A NEW APPROACH TO the education of nurses and allied professions. Each volume has been written by a leading expert who is in close touch with the education of nurses.

These books do not cover any particular examination syllabus but each one contains more than enough information to enable the student to pass his or her examinations in that subject. The aim is rather to provide the understanding which will enable each person to get the most out of and put the most into his or her profession. Throughout we have tried to present medical science in a clear, concise and logical way. All the authors have endeavoured to ensure that students will truly understand the various concepts instead of having to memorize a mass of ill-digested facts. The message of this new series is that medicine is now moving away from the poorly understood dogmatism of not so very long ago. Many aspects of bodily function in health and disease can now be clearly and logically appreciated: what is required of the good nurse or paramedical worker is a thoughtful understanding and not a parrot-like memory.

Each volume is designed to be read in its own right. However, four titles: *Physics, Chemistry and Biology*; *Anatomy*; *Biochemistry, Endocrinology and Nutrition* and *Physiology* provide the foundations on which all the other books are based. The student who has read these four will get much more out of the other books which relate to clinical matters.

We hope that a feature of this series will be regular revision. Critical comments from readers will be much appreciated as these will help us to improve later editions.

1

Introduction

Biochemistry is the study of the chemistry of living organisms, of the ways in which food is used to serve all the many needs of the body. Biochemistry is closely connected with nutrition, the study of the types and amounts of various materials required in the diet. Biochemistry is also inextricably intertwined with endocrinology, the study of hormones, for most of the hormones exert their actions by altering the behaviour of chemical reactions within the body.

The central problem in biochemistry is that of the supply of energy. Energy is needed for a multitude of purposes of which muscular activity is the best known. Energy is required for digestion, and for the functioning of the kidney, the liver, the brain and all the other organs in the body. Energy is also essential for the building up of the complex organic molecules of which the body is constructed.

Ultimately, most of the energy utilized on earth comes from the sun. Plants are able to tap this energy source directly by the process of photosynthesis. By using pigments, notably the green chlorophyll, plants can trap the energy of sunlight and use it to build up complex substances such as fat, carbohydrate, protein and nucleic acids. The only raw materials required are carbon dioxide, water and simple inorganic substances such as nitrates which can be extracted from the soil. In contrast to plants, animals lack pigments like chlorophyll and so they cannot directly trap the energy of sunlight and turn it to their own use. Instead they must obtain their energy indirectly by eating plants or other animals which have themselves eaten plants. The complex fats, proteins and carbohydrates which plant and animal tissues contain can then be broken down and made to yield up their energy.

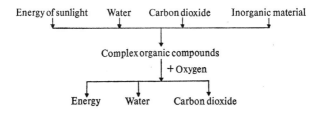

Fig. 1.1. The link between the biochemistry of plants and the biochemistry of animals.

The importance of ATP

Energy can be released by the breakdown of many different chemical compounds. This energy can then be used for the manufacture of other compounds, for muscular activity and for many other different purposes. How is the link between the release of energy and the employment of that energy established?

It may be helpful to consider the supply of energy in man-made systems. Undoubtedly electricity is by far the most important type of energy used by man. Electricity can be manufactured by a number of different processes. The energy of falling water, of coal, of wood, of oil, of natural gas and of the atom can all be converted by appropriate power generating stations into electricity. In turn, this electricity can be used by both domestic and industrial consumers in a myriad different ways, for lighting, for heating, for refrigeration, for cooking, and for driving the electric motors of thousands of different types of machine ranging from hair dryers to railway engines. Electricity is thus a most useful type of energy: it can be manufactured in many different ways and once made it can be used in many different ways.

Is there anything comparable to electricity in the body? There is. It is the substance called adenosine triphosphate or more familiarly as ATP. Like electricity it can be made in many different ways, notably by the breakdown of the carbohydrates, fats and proteins found in the food. Like electricity, the energy of ATP can be used in many different ways. It is employed in muscular contraction, in nervous activity and in the manufacture from simple substances of all the complex materials which the body requires. ATP is an almost universal substance. It is not confined to humans or even to mammals. It is the energy source which drives Olympic athletes, which lights up the tail of the firefly and which provides the electric shock of the electric eel. It is the central point in the chemistry of the body.

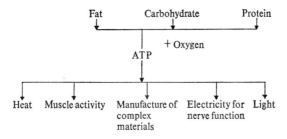

Fig. 1.2. ATP is the central substance in biochemistry. It can be produced by the breakdown of any foodstuff and used to supply energy for any type of activity.

Some important concepts and definitions

Metabolism is the term given to the sum of all the chemical reactions occurring at any time in the body. Anabolism indicates those reactions in which larger molecules are being built up from smaller ones: anabolism is a process which uses up ATP. Catabolism indicates those reactions in which large molecules are being broken down into smaller units: catabolism is the process which manufactures ATP.

Most of the ATP in the body is obtained by the breakdown of fats, carbohydrates and proteins in the presence of oxygen. This process is known as oxidation and its main end-products are carbon dioxide and water. It is similar in principle to the process of burning which also involves the breakdown of complex materials in the presence of oxygen. But whereas burning is usually uncontrolled and the energy released by it is wasted, the process of oxidation of food in the body is carefully regulated and much of the energy released is trapped in the form of ATP. The breakdown of food materials in the presence of oxygen is known as aerobic catabolism. Very much smaller amounts of ATP are formed by the breakdown of food in the absence of oxygen. This process is known as anaerobic (without air) catabolism.

It is useful to have some measure of the total amount of energy being used by the body. The most convenient way of expressing this is in the form of calories. When food is burned it gives out the energy stored in it in the form of heat. The amount of heat released is a measure of the amount of energy made available to the body when the same amount of food is oxidized in the process of catabolism. The heat output of the food is measured as calories. One calorie is the amount of heat required to raise 1 cm^3 of water through $1 \, °\text{C}$. The calorie itself is a very small unit and so the kilocalorie, which is 1,000 calories, is more often used. In older books, calorie with a small c means calorie while Calorie with a big C means kilocalorie. This has led to so much confusion that it seems wiser to use the scientific words calorie and kilocalorie. Therefore, when you read that such a quantity of such a food has a calorific value of 100 kcal, it means that that amount of that food when burnt in the presence of oxygen will release 100 kcal of heat. When oxidized in the body it will provide an amount of energy equivalent to 100 kcal of heat. This question of calories will be further discussed in chapter 9.

2

Important substances in biochemistry

This chapter outlines the role of various important substances in biochemistry. It is intended to give a bird's eye view of the subject so that when you come to later chapters you can see where each piece of information fits into the overall pattern.

Water

This is the single most important substance in the body. In a lean adult male, 60 per cent of the body weight is made up of water. In females and fat males the proportion of water is rather less. This is because fatty tissue contains very little water and so adding fat to the body adds very little water: the percentage of the body weight made up of water therefore falls. Water is important because most

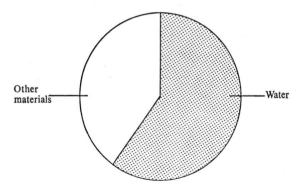

Fig. 2.1. In a lean male, 60 per cent of the body weight is water.

of the substances found within cells are dissolved in it. The blood which carries materials all around the body consists largely of water. The urine contains waste materials dissolved in water. These are just a few important examples of how important water is. It is by far the most important substance in the diet. When deprived of any other food material, any healthy individual can survive for weeks or even months. But if totally deprived of water, the healthiest of men will be dead within a few days.

Oxygen

When substances burn in a fire, they do so because they combine with oxygen. This combination leads to the energy stored in the substances being liberated as heat. The body too uses oxygen to 'burn' food materials but it does so in a much more controlled way than in a fire. Only very small amounts of energy are released in the body by catabolism in the absence of oxygen (anaerobic metabolism).

When oxygen is used (aerobic metabolism), the ultimate end products are water and carbon dioxide, in addition, of course, to ATP. The water may amount to 200–300 ml/day and should not be forgotten when calculating how much water a patient has received or lost.

Oxygen is supplied to the body via the lungs where the air comes into close contact with the blood. Carbon dioxide leaves via the lungs, moving in the reverse direction to oxygen.

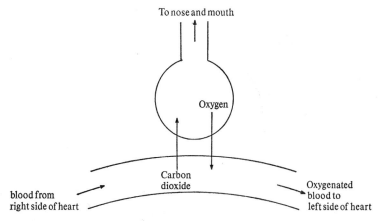

Fig. 2.2. The function of the lungs.

Carbohydrates

Carbohydrates are substances which contain carbon, hydrogen and oxygen in the ratio 1 : 2 : 1. For example, the formula of glucose, one of the best known carbohydrates, is $C_6H_{12}O_6$, i.e. one molecule of glucose contains six carbon atoms, twelve hydrogen atoms and six oxygen atoms. The basic building blocks of which carbohydrates

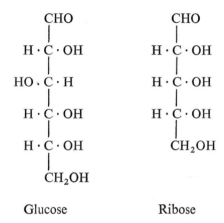

Glucose   Ribose

Fig. 2.3. The chemical formulae of glucose, a typical hexose monosaccharide with six carbon atoms; and of ribose, a typical pentose with five carbon atoms.

are made are known as monosaccharides. Monosaccharides with five carbon atoms are known as pentoses and those with six carbon atoms (like glucose) are known as hexoses. The hexoses are particularly important in metabolism: they include glucose, galactose and fructose. Pentoses are also important since two of them, ribose and deoxyribose, are essential components of the nucleic acids (see chapter 6).

Disaccharides are also both familiar and important. As their name suggests they contain two monosaccharide molecules linked together. Sucrose (cane sugar) contains a glucose molecule and a fructose molecule. Lactose, the main carbohydrate in milk, consists of a glucose and a galactose molecule. Maltose, frequently formed in the gut by the breakdown of more complex carbohydrates, consists of two linked glucose molecules.

Polysaccharides are made up of long chains of monosaccharide molecules. Only three are of much importance. Cellulose is the tough substance of which plant cell walls are made. It cannot be satisfactorily digested by man or by meat-eating mammals. It can be used only by animals which have special digestive structures such as the rumen of the cow. Such plant-eating animals have many bacteria in their guts and it is the bacteria which initially break down

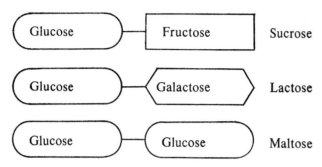

Fig. 2.4. The constitutions of sucrose, lactose and maltose.

the cellulose. Only when the bacteria have done their work can the animal absorb the breakdown products. Starch is another polysaccharide which is a plant product. It is the form in which carbohydrate is stored inside plant cells. Both starch and cellulose consist entirely of chains of glucose molecules. Finally, glycogen is the animal equivalent of starch. It is stored in cells in liver and muscle. It too consists solely of glucose. The differences between cellulose, starch and glycogen are due to differences in the ways in which the glucose molecules are strung together.

Fats

Fats too contain only carbon, hydrogen and oxygen. However, they contain less oxygen and hydrogen for each carbon molecule than do the carbohydrates. The fats are very much a mixed group of substances all of which share the property of being soluble in such substances as chloroform and ether. Some fats can dissolve in water as well but most tend to be water insoluble. Perhaps the most important groups of fats in the body are:

1. GLYCERIDES. Glycerides are built up by the combination of fatty acids with glycerol (popularly known as glycerine). Glycerol can combine with fatty acids by means of its three hydroxyl (OH) groups.

When one of the hydroxyl groups is combined with a fatty acid, the resulting compound is said to be a monoglyceride. When two hydroxyl groups are combined with fatty acids the result is a diglyceride and when three are combined the result is a triglyceride. The fatty acids themselves (sometimes known as free fatty acids, FFA, or non-esterified fatty acids, NEFA) consist of chains of carbon molecules with hydrogen ions attached to them and a carboxylic acid group

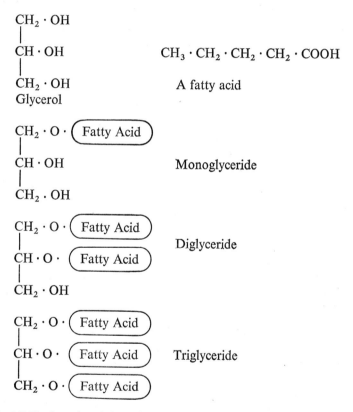

Fig. 2.5. The formulae of glycerol, a fatty acid and the various types of glycerides.

(COOH) at one end. They come in two main forms known as saturated and unsaturated. This may seem academic but it may turn out to be very important in practice. One theory which attempts to account for the present high incidence of heart disease in developed societies suggests that it is caused by a change in the balance of

B

saturated and unsaturated fatty acids in the diet. With saturated fatty acids, each carbon atom has two hydrogen atoms attached to it and so all the four valency bonds of carbon are used up, hence the word 'saturated'. With unsaturated acids, some carbons have only one hydrogen atom attached to them. There is thus a spare valency bond which is usually attached to the next carbon atom giving a double bond. Unsaturated fatty acids tend to be found in liquid fats such as palm, corn and soya-bean oils. Saturated acids predominate in solid fats. Margarine, for example, is made from oils. During its manufacture these oils are made to react with more hydrogen. This makes them more saturated and hence more solid.

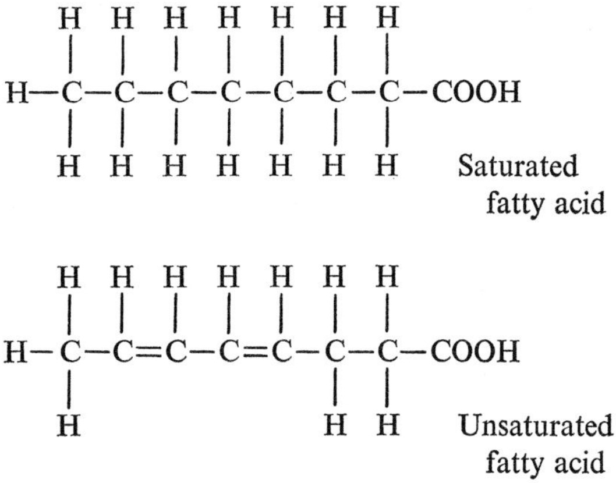

$$
\begin{array}{c}
\quad H \quad H \quad H \quad H \quad H \quad H \quad H \\
\quad | \quad\; | \quad\; | \quad\; | \quad\; | \quad\; | \quad\; | \\
H-C-C-C-C-C-C-C-COOH \\
\quad | \quad\; | \quad\; | \quad\; | \quad\; | \quad\; | \quad\; | \\
\quad H \quad H \quad H \quad H \quad H \quad H \quad H \qquad \text{Saturated} \\
\qquad\qquad\qquad\qquad\qquad\qquad\qquad \text{fatty acid}
\end{array}
$$

$$
\begin{array}{c}
\quad H \quad H \quad H \quad H \quad H \quad H \quad H \\
\quad | \quad\; | \quad\; | \quad\; | \quad\; | \quad\; | \quad\; | \\
H-C-C=C-C=C-C-C-COOH \\
\quad | \qquad\qquad\qquad\qquad | \quad\; | \\
\quad H \qquad\qquad\qquad\quad H \quad H \qquad \text{Unsaturated} \\
\qquad\qquad\qquad\qquad\qquad\qquad\qquad \text{fatty acid}
\end{array}
$$

Fig. 2.6. Formulae of a saturated and an unsaturated fatty acid to show the difference between the two.

The main function of the glycerides in the body is to act as a supply of energy. Some of the unsaturated fatty acids, known as essential fatty acids, have a vitamin-like function (see chapter 4).

2. STEROIDS. These are substances which have four rings of carbon atoms arranged in the form shown in fig. 2.7. Cholesterol is perhaps the best known steroid as it is found in the blood and high blood levels appear to be associated with heart disease. The hormones produced by the adrenal cortex and by the gonads (ovaries and testes) are also steroids.

3. PHOSPHOLIPIDS. These are complex substances whose functions are as yet poorly understood. As their name suggests they contain

phosphate groups. They seem to be important in the transport of fats in the blood, in the manufacture of cell membranes and in the structure of nervous tissue.

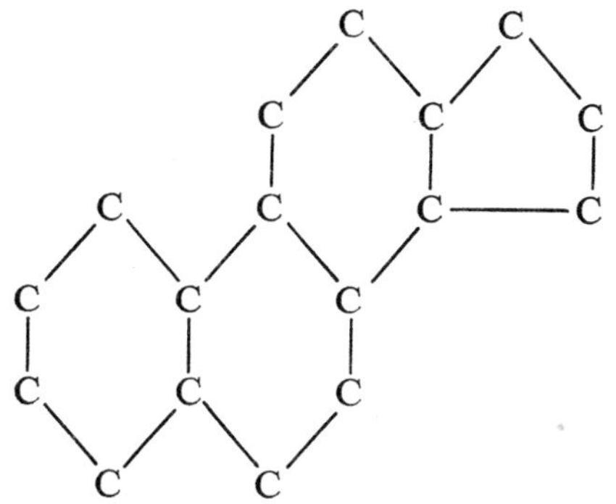

Fig. 2.7. The steroid nucleus on which is based the structure of cholesterol, of the adrenal cortical hormones and of the sex hormones.

Proteins

Proteins consist mainly of carbon, hydrogen, oxygen and nitrogen with much smaller quantities of sulphur. The large protein molecules are built up from smaller substances known as amino acids, of which about twenty types are important in human biochemistry. The amino acids are substances which carry both carboxylic (COOH) and amino (NH_2) groups.

Amino acids can link together in chains, with the carboxylic acid

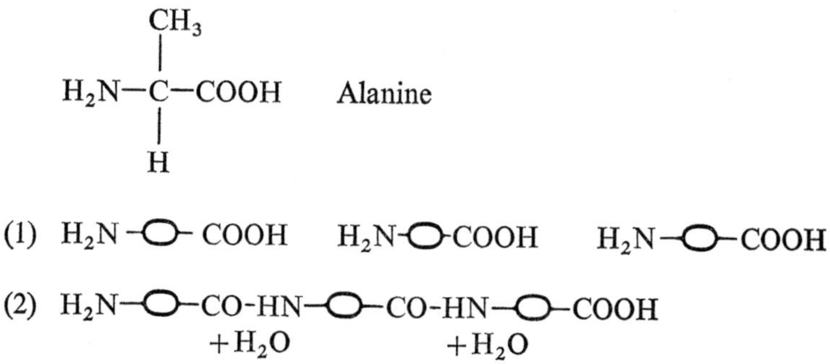

Fig. 2.8. Alanine is a typical amino acid. (1) Shows three separate amino acids. (2) Shows the three linked together by peptide bonds to form a short peptide chain. Water is eliminated in the course of the reaction.

group of one amino acid joined to the amino group of the next amino acid. Such chains are known as peptides and the links between the amino acids are known as peptide bonds. Proteins may contain one, two, three or four peptide chains. In the full protein molecule, the chains are folded and coiled up upon themselves in complex ways.

The proteins in the body are important in two main ways. They make up parts of the structure in many organs (e.g. the tendons of muscles). They also act as enzymes which are important in virtually all biochemical reactions.

Nucleic acids

These have been very much studied in recent years. They are important because the genetic material in the chromosomes consists of nucleic acids. Similarly, the ribosomes, structures in the cytoplasm of cells where most protein synthesis seems to take place are rich in nucleic acids. The nucleic acids are the vital materials which hand on inherited characters from generation to generation. They are also

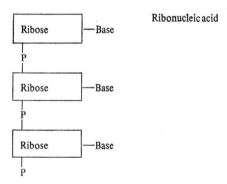

Ribonucleic acid

Fig. 2.9. The building blocks, ribose, phosphate and bases, of which ribonucleic acid is made.

important in directing the day to day life of each cell: disorders of the control which they normally exert over the cell's behaviour may be important in the causation of cancer.

Nucleic acids too are made up of simpler building blocks. Each molecule contains sugars, phosphate groups and complex organic materials known as bases. The nucleic acids are divided into two great groups depending on whether they contain the pentose sugar, ribose, or the slightly different pentose, deoxyribose. The former are known as ribonucleic acids (RNA's) and the latter as deoxyribonucleic acids (DNA's). The bases are shown in table 2.1. DNA is found primarily in cell nuclei, while RNA is found primarily in the cytoplasm. Both types of nucleic acid consist of chains of pentose molecules joined to one another by phosphate groups. The bases are attached to the sugars. In DNA two such chains are wound around one another in a helical formation (the famous 'double helix').

Table 2.1 *The bases found in the nucleic acids*

DNA	RNA
Adenine (A)	Adenine (A)
Thymine (T)	Uracil (U)
Guanine (G)	Guanine (G)
Cytosine (C)	Cytosine (C)

Enzymes

Enzymes are the protein molecules on which the whole of metabolism depends. Almost all the chemical reactions which occur in the body could take place without enzymes but they would take place far too slowly to have any biological value. Enzymes speed up and direct virtually all the chemical reactions of metabolism. The main characteristics of the enzymes themselves are:

1. They are all protein in nature but many of them have some other substance such as a metallic ion attached to the protein molecule.

2. Most enzymes are specialized to carry out a particular chemical reaction.

3. All enzymes require special environmental conditions if they are to function normally.

The specialization of function occurs because in order to work, an enzyme must actually combine with the substance which it is chemically changing. Substances acted upon by enzymes are known as substrates. A combination between an enzyme and its substrate may be likened to a lock and key mechanism. The enzyme is the lock: because it is of a particular chemical structure, only substrates whose chemical structure can fit the lock can become attached to the enzyme (fig. 2.10). Only when the substrate has become attached to the enzyme can the enzyme function. Enzymes function in three main ways:

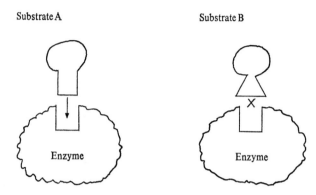

Fig. 2.10. The nature of enzyme action. The substrate is like a key which must be able to fit into a slot on the enzyme molecule before the enzyme can act. In this case the enzyme can act on substrate A but not on substrate B.

1. They may break the substrate molecule into two parts. This happens particularly in digestion.

2. They may break off a small part of the substrate molecule, usually a hydrogen atom or a CH_3 group, or a water molecule or a carbon dioxide molecule.

3. They may build up the substrate molecule by adding a small part to it.

Whatever happens, once the change has been brought about, the substrate is released from combination with the enzyme and the enzyme is ready to repeat the process on another substrate molecule. From this brief description the following important points about enzyme behaviour may be deduced:

1. Most enzymes are highly specialized so that they can react with only one type of substrate.

2. Each enzyme carries out only one process. Therefore in order to break down a large molecule, or in order to build up a large one from much smaller units, many enzymes are required, each performing one stage in the whole process. The enzymes do not swim around inside cells in a form of soup. Instead they are attached in

sequence to membranes. When a substrate molecule has been dealt with by one enzyme, the substrate is passed immediately to the enzyme which carries out the next stage in the process and which is attached to the membrane immediately adjacent to the first enzyme.

3. There is a limit to the rate at which a single enzyme molecule can work. If there is an excess of substrate available then all the enzyme molecules will be working at maximum capacity and the more enzyme molecules there are the faster will the substrate be changed. If all the enzyme molecules are not working maximally, then the rate of change of the substrate depends on how rapidly the substrate can be supplied to the enzymes.

4. Many enzymes act by either adding to or removing from substrates atoms or small groups of atoms. If the former type are to work effectively they must be constantly supplied with the atoms and small groups which they use. If the latter are to function properly the atoms and small groups must be continuously removed or they will accumulate and gum up the works. The supply and removal

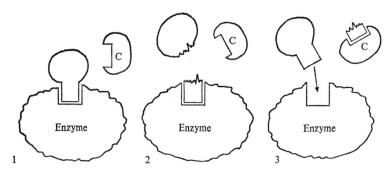

Fig. 2.11. The action of a coenzyme. The enzyme breaks a small piece off the substrate which is then carried away by the coenzyme so leaving the enzyme free to act again.

of these atoms and small groups is the work of substances known as coenzymes. Many of the vitamins, particularly those of the B group, appear to be so important because they function as coenzymes in vital reactions.

Protein molecules are very easily damaged and if they are to function properly, enzymes require very special conditions. The temperature must be kept within very narrow limits: if it rises too high or falls too low, the rate at which the enzymes work is altered and the enzyme molecules themselves may be destroyed (or denatured as the process is sometimes known). Most enzymes in mammals work best at about the temperature of the blood and this is a major reason why the temperature of the blood and of the body must be kept constant. The pH of the body fluids must also be kept more or less constant. The term pH is a useful shorthand way of describing acidity or alkalinity. Acid solutions have a low pH and alkaline solutions have a high one. Most enzymes again work best at the pH of the blood. There are some exceptions to this rule, however, for example: the enzymes in the stomach work best in

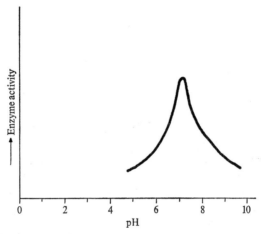

Fig. 2.12. The influence of pH on enzyme activity. Most enzymes in the body have peak activity around pH 7.

quite strong acid and cease to function at the pH of the blood. Finally the osmotic pressure and chemical composition of the body fluids must also be kept constant if the enzymes are to work normally. This maintenance of a constant chemical composition is primarily the job of the kidney.

Enzymes are now named by adding '-ase' to the name of the substrate on which the enzyme acts. For example, enzymes which break down proteins are called proteases, those which break down lipids are called lipases and so on. Some of the enzymes which were discovered many years ago were named before this system was introduced. These enzymes usually keep their own names, like the pepsin which digests proteins in the stomach and the trypsin of pancreatic juice which digests proteins in the intestine.

Hormones

Glands are organs which manufacture and secrete chemical substances. They can be divided into two main types. Glands with ducts discharge their secretions down the duct until it reaches some sur-

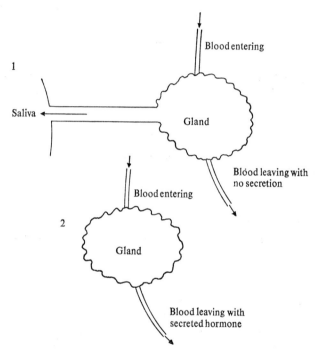

Fig. 2.13. The differences between glands with ducts like the salivary glands (1) and glands without ducts which secrete hormones into the blood (2).

face: the salivary glands secrete saliva down the salivary ducts until it reaches the surface of the mouth: the sweat glands secrete sweat along ducts onto the surface of the skin. Ductless glands, in contrast, do not secrete their products into ducts which discharge at a specific point. Instead they secrete their products into the blood which passes through the gland. The blood then carries the chemicals to every region of the body. The substances secreted in this way are known as hormones and the glands themselves as ductless or endocrine glands. For the most part the hormones exert their actions by altering the working of metabolism in some way.

Outline of catabolism

Carbohydrates, fats and proteins are all capable of being broken down to supply energy. Although these substances are so different, the final and most important stages of their destruction are shared with one another. The first aim of catabolism is to break down,

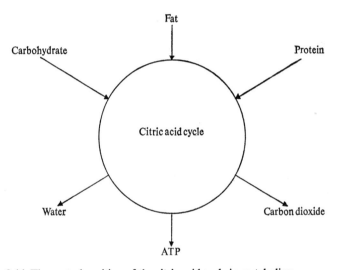

Fig. 2.14. The central position of the citric acid cycle in metabolism.

usually without the aid of oxygen, the large complex molecules into smaller ones. The small molecules resulting from the breakdown of all three foodstuffs can then enter the series of reactions known as the citric acid, Krebs, or tricarboxylic acid cycle. Once in the cycle they are further broken down with the aid of oxygen to give water, carbon dioxide and large amounts of ATP.

3

Carbohydrate metabolism

Carbohydrates occur in human food in three particularly important forms, as the polysaccharide starch, and as the disaccharides sucrose (cane sugar) and lactose (milk sugar). All three must first be broken down to the monosaccharides glucose, fructose and galactose before they can be absorbed from the gut into the blood. This process is known as digestion and the main stages are:

1. Saliva contains an enzyme known as amylase or ptyalin, which can break down starch into smaller units. The process does not continue very far as amylase is inactivated by the acid gastric juice.

2. Gastric juice contains no enzyme which breaks down carbohydrates. However, the acid itself may help to break up the tough starch granules, so making them more easily digested in the intestine.

3. The pancreatic juice which is poured into the duodenum also contains amylase which completes the breakdown of starch to the disaccharide maltose. The bicarbonate in the pancreatic juice neutralizes the gastric acid and so allows the enzyme to work.

4. The wall of the intestine itself manufactures enzymes which break down the disaccharides to monosaccharides. Maltase breaks down

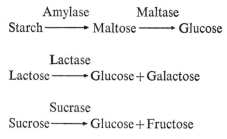

Fig. 3.1. The digestion of the main carbohydrates.

maltose to glucose. Lactase breaks down lactose to glucose and galactose. Sucrase (sometimes known as invertase) breaks down sucrose to glucose and fructose. If one of these enzymes is absent at birth, the infant will be unable to digest one of the disaccharides (most commonly lactose). The undigested food passes through the gut causing diarrhoea: ultimately malnutrition will occur because the baby cannot obtain enough calories. A remarkable improvement occurs once the offending disaccharide is removed from the diet.

The portal vein and the liver

The monosaccharides, glucose, galactose and fructose, pass through the intestinal wall into the blood of the hepatic portal system. The hepatic portal vein carries all the blood from the gut to the liver.

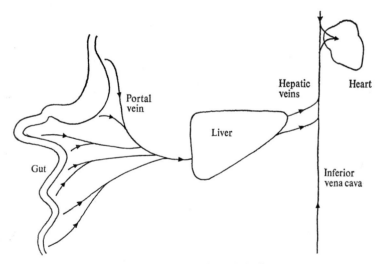

Fig. 3.2. The arrangement of the portal vein and the liver.

Only after passing through the liver does the blood from the gut mix with the blood from the rest of the body returning to the right side of the heart.

In the liver, much of the glucose is removed from the blood and is built up into the polysaccharide, glycogen, in which form it is stored. Much of the galactose and fructose is also removed from the blood and liver enzyme systems convert these substances into glucose, the central substance in carbohydrate metabolism. These converting enzyme systems may also be congenitally defective. If they are, either galactose or fructose may not be converted to glucose. A deficiency of the galactose enzymes is not uncommon and is

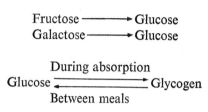

Fig. 3.3. The treatment of digested carbohydrates by the liver.

particularly dangerous because of the large amounts of galactose formed from the lactose of milk. Galactose accumulates in the blood in large amounts (galactosaemia) and may damage the brain. Once the condition is recognized and lactose is removed from the diet, a rapid improvement occurs.

Between meals, when no glucose is entering the blood from the gut, the glycogen in the liver is slowly broken down. The glucose produced in this way is released into the blood in order to keep the level of blood glucose constant. It is important that glucose concentrations in the blood should be kept constant because glucose can be used as an energy source by virtually every cell in the body. The constancy is particularly vital for the function of the brain. Most other cells in the body can get their energy from carbohydrate or fat but this does not seem to apply to the nerve cells. The brain seems to rely entirely on glucose for its energy supply and it is incapable of using fat or protein. Therefore if the blood glucose concentration becomes too low, the brain cells may be permanently damaged.

What happens to blood glucose?

Most of the glucose in the blood is removed from it and immediately used by cells to provide energy. Some is taken up by the muscles and heart for storage in the form of glycogen in order to tide over possible emergencies. Some is taken up by fatty tissue and converted into fat, another form of food storage.

The direct breakdown of glucose by cells can be divided into two stages, anaerobic in which no oxygen is required, and aerobic in which oxygen is essential. In the first of these stages, glucose is broken down to pyruvic acid: in the absence of oxygen, the process can go no further and the pyruvic acid is converted to lactic acid, a blind alley in metabolism. In humans and other mammals oxygen is

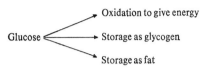

Fig. 3.4. The fates of blood glucose.

usually available and lactic acid is formed primarily during muscular exercise when the oxygen supplies reaching the muscle may not be quite sufficient. However, many micro-organisms, such as yeasts, may not have the enzymes to carry the breakdown of pyruvic acid further. They, like humans, convert glucose to pyruvic acid but then they convert the pyruvic acid to ethyl alcohol: this reaction is the foundation of the whole brewing and wine making industry.

In exercising humans, some of the lactic acid produced stays in the muscles while some escapes into the blood and reaches the liver. When the exercise stops and more oxygen is again available, the lactic acid is converted back to pyruvic acid which is either oxidized completely or converted back to glucose and glycogen. Up to the stage of pyruvic acid, only very small amounts of ATP are produced from each glucose molecule.

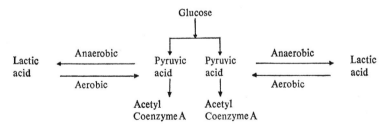

Fig. 3.5. The breakdown of glucose to acetyl coenzyme A. Each molecule of glucose breaks in half and yields two molecules of acetyl coenzyme A.

Aerobic metabolism

When ample oxygen is available, the pyruvic acid is converted into a very important substance known as acetyl coenzyme A. This conversion is particularly interesting because it demonstrates the part which vitamins can play in the body. Four vitamins, thiamine, lipoic acid, pantothenic acid and nicotinic acid are essential for the reaction to occur.

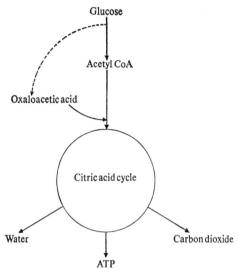

Fig. 3.6. The importance of oxaloacetic acid in the entry of acetyl coenzyme A into the citric acid cycle.

Acetyl coenzyme A is important because it is the means whereby carbohydrates can enter the citric acid cycle which completely oxidizes them to give water, carbon dioxide and large amounts of ATP. The conversion of one glucose molecule to pyruvic acid yields only four molecules of ATP, while the conversion of the two molecules of acetyl coenzyme A formed from one glucose molecule to carbon dioxide and water yields thirty-four molecules of ATP. It is obvious that aerobic metabolism is a much more effective and important process.

Finally, as will become apparent later when discussing diabetes, it is important to note that acetyl coenzyme A enters the citric acid cycle by combining with a substance known as oxaloacetic acid. Oxaloacetic acid too can be formed from glucose via a side path of metabolism.

4

Fat metabolism

Fats are almost as important as carbohydrates in the supply of the energy needs of the body. Most of the fat in the food is in the form of triglycerides. Fats differ from carbohydrates in that although they must be partially digested, they do not need to be broken down completely into their basic building blocks of glycerol and fatty acids before they can be absorbed across the gut wall.

No digestion of fat takes place in the saliva and little if any in the stomach. However, in the stomach the fat is softened and possibly small amounts may be broken down by the direct action of the stomach acid. But there can be no doubt that the major part of fat digestion begins when the food leaves the stomach and enters the first part of the small intestine, the duodenum. There it meets with two important juices, the bile secreted by the liver and the pancreatic juice secreted by the pancreas.

The bile (see chapter 7) does not actually digest the fat. It contains no fat-breaking enzymes (lipases) but it does contain the bile salts of glycocholic and taurocholic acids. The bile salts are detergents, similar in many ways to the detergents which are used for washing up dirty dishes. The first problem of fat digestion is how can the lipases which are found in the pancreatic juice get at the fat? The enzymes are secreted in the form of a watery solution which cannot penetrate into the fat globules of the food just as ordinary water cannot break up the large fat globules in the frying pan. The enzymes can only get at the fat molecules if the large, unwieldy fat globules are broken down into tiny droplets. This breaking up of the globules is performed by means of the detergents in the bile and is similar in principle to the clearing of the frying pan fat by washing-up fluid. The process is known as emulsification. Once the fat is in the form of tiny droplets, then it is much more accessible to the enzymes of the pancreatic juice.

The pancreatic lipases break off fatty acids from the triglycerides. The droplets become still smaller and consist of a complex mixture of glycerol, free fatty acids, monoglycerides, disaccharides, triglycerides and bile salts. These droplets can then pass through the gut wall even without the complete digestion of triglycerides to give glycerol and free fatty acids. Although the droplets can pass through the gut wall they are still too large to cross the walls of the capillaries to enter the blood. Instead they enter the much more porous lymphatic vessels (see *Physiology* text). They are carried by the lymph directly into the venous blood, by-passing the portal system and the liver.

Fat catabolism

There are three main stages in the breakdown of fats within the body. First of all the glycerides are split up to give glycerol and free fatty acids. Glycerol is a substance which is part way between glucose and pyruvic acid in structure and so it can easily be broken down by the enzymes which metabolize glucose. Fatty acids are quite different. Two carbon atoms at a time are split off from the chain of carbon atoms in the fatty acid molecule. These two-carbon fragments are converted to acetyl coenzyme A. Finally the acetyl coenzyme A

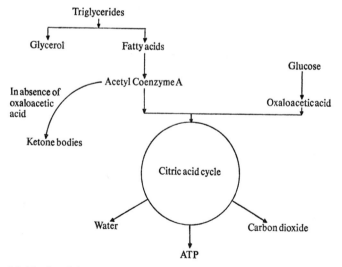

Fig. 4.1. The breakdown of fats.

C

combines with oxaloacetic acid and is completely oxidized by the citric acid cycle. Thus both fats and carbohydrates are finally broken down by the same route.

It is worth noting that both fats and carbohydrates must enter the citric acid cycle by combining with oxaloacetic acid. Oxaloacetic acid is manufactured from carbohydrate. This means that carbohydrates can supply their own oxaloacetic acid and do not need any fat oxidation for their complete breakdown. In contrast the fats cannot supply their own oxaloacetic acid and hence they cannot be fully oxidized unless some carbohydrate is also available.

A deficiency of carbohydrate is most likely to occur in two situations. In starvation, the carbohydrates are rapidly used up and the body must rely on its fat stores for its energy supply. In diabetes mellitus there may be plenty of carbohydrate available but it cannot be used normally because of a lack of insulin (see chapter 13). In these circumstances, because of lack of oxaloacetic acid, fat breakdown tends to be halted at the stage of acetyl coenzyme A. Acetyl coenzyme A molecules therefore accumulate and react with one another to give three other substances, acetoacetic acid, beta-hydroxybutyric acid and acetone. These three substances are known as ketone bodies and when they accumulate the patient is said to be suffering from ketosis. The whole problem will be further discussed in chapter 13.

Fat storage

Carbohydrate is stored in small quantities in liver and muscles in the form of glycogen. These glycogen stores can supply energy needs for only 24–48 hours after which they are completely used up. The main energy stores in the body are in the form of fat. This seems to be because a given weight of fat yields considerably more energy than the same weight of carbohydrate. It is therefore more efficient to store energy in the form of fat.

The fat is stored in special cells in the tissue known as adipose tissue. Adipose tissue is mainly found immediately beneath the skin, but there are often large amounts in the peritoneal cavity in the folds of the omentum. Under the skin, the breasts, buttocks and abdominal wall seem to be the sites most preferred. The fat in adipose tissue is found in the form of triglycerides. These may be built up directly from the fatty acids in the blood or they may be made indirectly from carbohydrate in the food not immediately required by the body.

When a person takes in more calories than he or she needs, the excess energy is stored in the form of fat. When more calories are used up than are supplied in the food, the extra calories needed are provided first by the breakdown of glycogen and then by the breakdown of fat.

Essential fatty acids

Fats are obviously of major importance in the supply of energy. But some types of fat have other roles as well. The steroids are hormones and the phospholipids are essential for the structure of cell membranes and of the central nervous system. The unsaturated essential fatty acids play a poorly understood role but one which may turn out to be of the utmost significance. The important essential fatty acids are called linoleic, linolenic and arachidonic acids. They are similar to vitamins in that they are required in very small quantities in the diet because they cannot be manufactured by the body from other substances. Their precise actions are uncertain but they appear to be essential for the manufacture of normal cell membranes. If experimental animals are completely deprived of essential fatty acids they develop a syndrome in which the skin and kidneys are damaged and reproduction fails. However, they are so widely distributed in the diet that these extreme conditions resulting from total lack are unlikely to occur naturally. Their main interest is that there is some evidence that a partial deficiency in humans may produce defective arterial walls and lead to the development of the very common disease, atherosclerosis. This is as yet unproven and is the subject of intensive research. There are in fact many theories as to the cause of atherosclerosis and not one is as yet well established.

5

Proteins

The provision of energy is the major function of both the fat and carbohydrate in the food. Both fat and carbohydrate contain only carbon, hydrogen and oxygen and can therefore be completely broken down to give carbon dioxide and water. Proteins differ in two main ways. First, their major function is to provide the amino acids from which enzymes and other important constituents of the body can be manufactured: they can be broken down to supply energy but this is a function of secondary importance. Secondly, in addition to carbon, hydrogen and oxygen they contain large amounts of nitrogen and small amounts of other substances such as sulphur. They cannot therefore be broken down to carbon dioxide and water alone since neither carbon dioxide nor water contains nitrogen. The nitrogen must be disposed of somehow. This is primarily the task of the kidneys which excrete most of it in the form of urea with smaller amounts as uric acid and ammonium ions.

Digestion

The proteins taken in with the food cannot be used directly by the body. Before they can be absorbed into the blood they must be broken down in the gut to give their constituent amino acids.

The first important stage of protein digestion takes place in the stomach. There an inactive enzyme, pepsinogen, is secreted by glands in the stomach wall. It is important that this protein-digesting enzyme should be secreted in an inactive form, otherwise it would digest the cells in which it was made. On entering the stomach it comes into contact with the acid gastric juice which is secreted by other cells in the stomach wall. The acid breaks off a small part of the inactive pepsinogen molecule and converts it to active pepsin.

The pepsin can then begin to break up the amino acid chains of which the protein molecules are made. It works best at a pH of 2–3 which is the highly acid pH normally found in the stomach. The digestion of proteins to the level of single amino acid units is completed in the intestine. Protein-splitting enzymes are found both

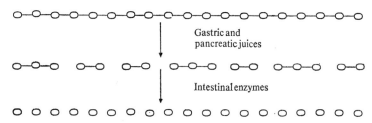

Fig 5.1. The digestion of protein. Each oval represents an amino acid.

in the juice secreted by the pancreas and in the secretions of glands in the wall of the small intestine itself. All these enzymes work best when the pH is nearly neutral (around 7). The first stage therefore in protein digestion in the intestine is the neutralization of the acid material coming from the stomach by the large amounts of bicarbonate found in the pancreatic juice.

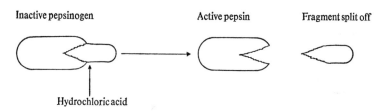

Fig. 5.2. The activation of pepsinogen by the hydrochloric acid in the stomach.

The pancreas secretes a number of inactive enzyme precursors of which the best known is trypsinogen. Again these substances must be made in an inactive form so that they do not digest the pancreas itself. In the disease pancreatitis, the enzymes are abnormally activated within the pancreas itself. They therefore digest the gland and escape into the abdominal cavity doing a great deal of damage and causing intense pain. Under normal circumstances, however, trypsinogen is not converted to active trypsin until it leaves the

pancreatic duct and enters the small intestine. There it meets a substance called enterokinase which is secreted by the intestine wall.
Enterokinase splits off a small part of the trypsinogen molecule
leaving the active enzyme trypsin. By means of both pancreatic and
intestinal enzymes, the proteins are broken down into the amino
acids which can cross the gut wall and enter the portal blood.

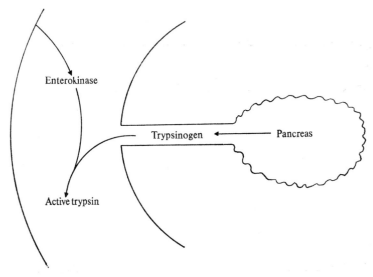

Fig. 5.3. The activation of trypsinogen by enterokinase in the intestine.

Essential amino acids

Twenty-one amino acids have so far been discovered in the proteins
found in mammals. Some of these cannot be manufactured in the
human body in sufficient amounts to supply the body's needs. They
must therefore be supplied in the food and are known as essential
amino acids (table 5.1). Protein is sometimes said to be of high
biological value when it contains all the essential amino acids and
of low biological value when one or more of the essential acids are
missing (see chapter 9).

The other amino acids are sometimes called non-essential. In some
ways this is misleading because they are in fact essential for the
body's economy and are vital to life. They are called non-essential
simply because they do not need to be supplied in the food: they

can be manufactured by the body itself. Surplus quantities of the amino acids the body does not immediately need can be used to manufacture those amino acids which it does need.

Table 5.1 *The essential and non-essential amino acids*

Essential	Nonessential
Arginine	Alanine
Histidine	Aspartic acid
Leucine	Cystine
Isoleucine	Glutamic acid
Lysine	Glycine
Methionine	Hydroxyproline
Phenylalanine	Proline
Threonine	Serine
Tryptophan	Tyrosine
Valine	Cysteine
	Hydroxylysine

Fates of amino acids

Four types of thing may happen to the amino acids which result from the digestion of protein in the food.

1. They may be used unchanged for the manufacture of new protein.

2. If there is an excess of a particular amino acid, that acid may be converted to another amino acid of which there is a shortage.

3. The amino acids may be broken down. The parts containing carbon, hydrogen and oxygen enter the citric acid cycle and are oxidized to supply energy. Thus the citric acid cycle is the final route of oxidation for all three major foodstuffs, fats, carbohydrates and

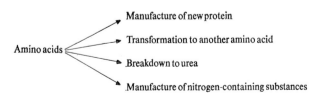

Fig. 5.4. The fates of amino acids.

proteins. The nitrogen-containing part of the amino acid molecule cannot be dealt with in this way. Instead it is usually converted to urea and excreted in the urine.

4. They may be used for the manufacture of non-protein nitrogen containing substances such as thyroid hormone, the nucleic acids and the haem part of haemoglobin.

The formation of urea

The first stage in the breakdown of the amino acids is the splitting off of the amino groups to give ammonia. This can occur in many organs but is most likely to take place in the liver. The process is known as deamination.

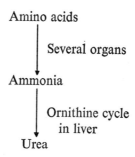

Fig. 5.5. The formation of urea from amino acids.

Ammonia is a highly toxic substance and if allowed to accumulate it may damage many organs and in particular the brain. It must therefore be converted as quickly as possible to something which is relatively harmless, highly soluble in water and excretable in the urine. The substance with these properties is urea and the conversion of ammonia to urea takes place in the liver. When ammonia is formed in other organs the blood rapidly takes it to the liver where it is rendered harmless. The sequence of reactions in the liver by which ammonia is converted to urea is known as the ornithine cycle.

6

Nucleic acids and protein synthesis

The chromosomes found in the nucleus of every cell are responsible for the development of that cell and for the control of the behaviour of the mature cell. The chromosomes seem to consist primarily of the nucleic acid known as deoxyribonucleic acid or DNA. The DNA carries the 'Plans' for the structure of every protein molecule manufactured by the cells. Since the proteins include the enzymes and since the enzymes are responsible for the manufacture of all non-protein substances in a cell, the DNA is in the end responsible for the manufacture of all substances made by each cell.

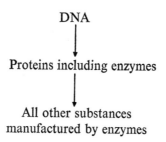

Fig. 6.1. DNA controls the manufacture of all the proteins including the enzymes. In turn the enzymes determine the manufacture of all other substances.

Cell development

The most important constituents of the human fertilized egg are the forty-six chromosomes, twenty-three derived from the father's sperm and twenty-three from the mother's ovum. These forty-six chromosomes carry the plans for the whole development of the body. Whenever a cell divides, exact copies of the forty-six chromosomes are made so that each of the pair of daughter cells contains a set of chromosomes identical to the set of the parent cell.

Several thousand different types of proteins are probably required for the normal working of the human body. Most of these are enzymes. Some are hormones such as insulin and growth hormone and others are substances serving a quite different function such as the globin part of the blood pigment haemoglobin. The DNA in the forty-six chromosomes carries the plans for the manufacture of all these proteins. The part of a chromosome which deals with the manufacture of one protein is known as a gene. It is possible for one gene and therefore one protein to be faulty even though the rest of the body is normal. For example, in haemophilia the gene which carries the instructions for the manufacture of the protein anti-haemophilic globulin (a clotting factor) is faulty. This means that the globulin cannot be manufactured normally and the blood does not clot normally. In galatosaemia (chapter 3) the gene which carries the instructions for one of the enzymes involved in the conversion of galactose to glucose is faulty and therefore galactose accumulates in the blood. Such genetic errors which affect the biochemistry of the body are sometimes known as inborn errors of metabolism.

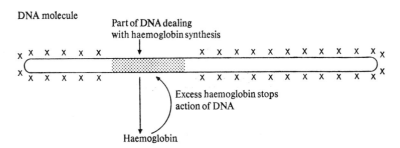

Fig. 6.2. The action of DNA in the red cell. Most of the DNA, except for the piece which carries the plans of haemoglobin, is inactivated. When sufficient haemoglobin accumulates in the cell this information is carried back to the DNA which then stops the manufacture of more haemoglobin.

Although the DNA in every cell in the body of any individual is identical and is therefore capable of stimulating the manufacture of every protein that the body can make, in practice each cell becomes specialized so that it manufactures only a few types of protein. Only a few genes are allowed to function and the activity of the others is more or less permanently suppressed. For example, red cells are specialized to manufacture haemoglobin: the parts of the DNA which carry the plans for insulin or for the enzymes used in the synthesis of fats are inactive. In contrast, the cells in the islets of Langerhans in the pancreas produce primarily insulin while the cells in adipose tissue produce primarily those enzymes required to deal with fats. Each cell is therefore specialized so that it produces only a limited number of the proteins which it could theoretically manufacture.

In addition, the manufacture of the proteins which each cell does produce is also strictly controlled. When a cell contains sufficient of a particular protein, information about this is sent back to the nucleus; the DNA stops further production of that protein. When the amount of that protein in the cell falls again, the DNA is again informed and this time it orders the manufacture of more of that protein. In this way the concentration of each type of protein within a cell is kept close to the desired level.

The genetic code

Each protein molecule consists primarily of a chain of amino acids. The properties of the protein molecule depend on the types of amino acids in the chain and on the order in which those amino acids occur. The DNA must therefore contain the plans for each protein molecule in terms of plans for a chain of amino acids. How does it do this?

In chapter 2 we saw that DNA contains four types of bases, adenine (A), thymine (T), guanine (G) and cytosine (C). It has been found that the sequence in which these bases occur on the DNA molecule determines the sequence in which the amino acids occur in the protein molecule. It has also been found that in order to specify one amino acid a sequence of three bases is required—the sequence of bases which specifies each amino acid is sometimes known as the genetic code. This may perhaps be best understood if specific examples are used. The amino acid phenylalanine is specified by the base code AAA, glutamic acid is specified by the base code CTC, while tyrosine is specified by the base code ATA. Thus the sequence of bases CTC–ATA–AAA on DNA would be the plan for a short peptide chain of composition glutamic acid–tyrosine–phenylalanine. Longer sequences of bases on DNA can specify the amino acid composition of any protein in the body.

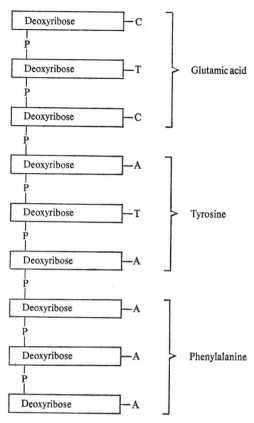

Fig. 6.3. A schematic diagram of a short length of a DNA molecule showing how the genetic code works.

Protein synthesis

The mechanism of protein synthesis has been worked out only recently. The bulk of such synthesis takes place in structures found in the cytoplasm of cells known as ribosomes. The ribosomes are largely made up of a special type of ribonucleic acid known as ribosomal RNA. The ribosomes are obviously well away from the DNA which is in the nucleus. A big problem in protein synthesis is therefore how the DNA gets its message across to the ribosomes. It may be helpful to consider the building of a protein molecule as similar in many ways to the building of a house. In house building, four categories of men are required:

1. An architect to draw up the plans.

2. A foreman to look at the plans and to supervise the actual construction of the house.

3. A gang of labourers to bring the necessary building materials to the site.

4. Another gang of labourers actually to put up the house.

In building a protein molecule the plans are provided by the DNA of the cell nucleus: it acts as the architect. The protein molecules are assembled from the amino acids by the ribosomes: they are the labourers who actually put up the building. The amino acids are brought to the ribosomes by another special type of RNA known as soluble or transfer RNA: the molecules of transfer RNA are the labourers who bring the materials to the site. Transfer RNA appears to float free in the cytoplasm. There are in fact over twenty different types of transfer RNA, each one specialized to combine with a particular amino acid. For example, there is a transfer RNA which combines with leucine, another which combines with alanine, another which combines with threonine and so on. The job of these transfer RNA molecules is to pick up the amino acids which reach the cell from the blood and to carry them to the ribosomes. There the link between the transfer RNA and the amino acid is broken. The amino acid is used for protein synthesis and the transfer RNA is free to go off to collect another amino acid molecule.

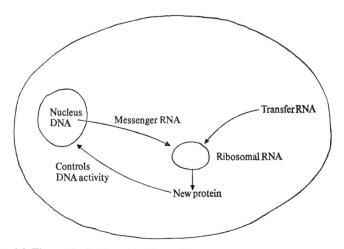

Fig. 6.4. The synthesis of proteins in a cell.

The remaining problem therefore is the link between nuclear
DNA and ribosomal RNA. What acts as the 'foreman' to instruct
the ribosomes how to carry out the plans of the nucleus? The link
is in fact made by yet another type of RNA, this time known appro-
priately enough as messenger RNA. Suppose that the nucleus

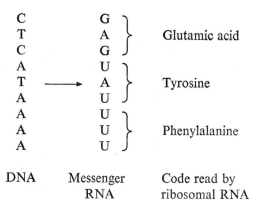

Fig. 6.5. The transfer of the coded message about protein structure from the
DNA to the ribosomes.

decides that a particular protein needs to be manufactured. That
protein consists of a chain of amino acids and the plan for it is
represented by a chain of groups of three bases on the DNA molecule. The first thing that happens is that an impression of this length
of DNA is made by manufacturing a messenger RNA molecule
which is complementary to it. This is possible because each base in
the DNA molecule orders the insertion of another base in the
messenger RNA molecule. For example, DNA adenine orders uracil
in messenger RNA, DNA thymine orders RNA adenine, DNA
guanine orders RNA cytosine and DNA cytosine orders RNA
guanine. Thus a sequence of nine DNA bases, CTC–ATA–AAA,
would become on the messenger RNA molecule, GAG–UAU–
UUU. Once the full messenger RNA molecule has been made in
this way it leaves the nucleus and moves out to a ribosome in the
cytoplasm. The ribosomal RNA seems to be capable of 'reading'
the code. For example, it 'sees' that the first three bases are GAG
and it 'knows' that this sequence means glutamic acid. It therefore
starts off the protein molecule by taking a molecule of glutamic acid
from the transfer RNA. It then 'reads' UAU and so a molecule of

tyrosine is taken from the cytoplasm and tagged on to the glutamic acid. UUU means phenylalanine and so the third amino acid is phenylalanine. The ribosome thus reads the whole chain of the messenger RNA molecule and builds up a corresponding chain of amino acids. On being completed the amino acid chain is released and spontaneously folds up to become the complete protein molecule.

Significance in disease

You may think that all this talk of DNA, RNA and protein synthesis is high-powered academic stuff which has little relevance to practical medicine. Nevertheless this modern work has greatly helped our understanding of two of the very common types of diseases with which nurses must deal. These are cancer and the virus-caused diseases.

We now know that most viruses consist of an outer protein coat with an inner core of nucleic acid. Most of the viruses which attack humans contain DNA although some contain RNA. The virus becomes attached to a cell by means of its protein coat. The protein then seems to bore a hole in the side of the cell and injects the

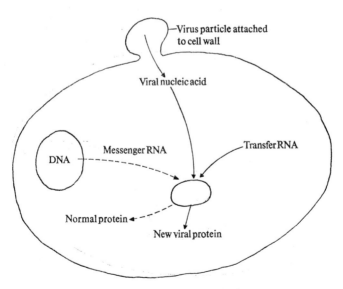

Fig. 6.6. The action of a virus on cell-protein synthesis.

nucleic acid into the cell interior. Once inside the nucleic acid can do two things. First it may manufacture its own equivalent of messenger RNA. This goes to the ribosomes which are then instructed to manufacture new virus particles instead of the normal cell proteins. The ribosomes are thus partially taken over by the invading virus. Secondly the virus seems to interfere with the normal controls over protein manufacture. When a particular protein accumulates in the cell, the DNA in the nucleus is no longer informed and so production of that protein continues and is not stopped as it usually would be. The cell must grow to accommodate the extra protein and quite soon it divides. The virus thus promotes abnormal cell multiplication. In some cases the virus seems to have primarily the first action. It takes over the ribosomes so completely that the cell ceases to manufacture its own protein and so soon dies. This seems to happen with a disease like smallpox. On the other hand, if the virus allows the cell to manufacture some of the cell's own protein, the cell will not die but will grow, divide and multiply in an uncontrolled way. This is what happens with the familiar warts which are caused by a virus. More seriously it occurs much more dramatically in the large numbers of tumours in animals which are now known to be caused by viruses. No human tumour has yet been shown to be virus-induced but most research workers now believe that some types of human cancer at least are caused by viral infection.

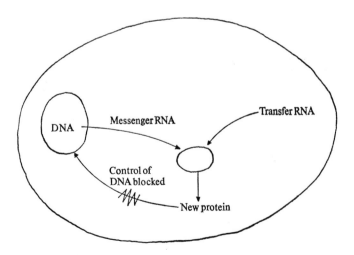

Fig. 6.7. The possible mechanism of cancer.

In all types of cancer, the normal controls which stop excess protein synthesis and cell multiplication appear to be lost. As the cell manufactures protein and that protein accumulates, the process is no longer stopped by information carried back to the DNA. The cell must divide and rapid multiplication takes place. Furthermore, many of the controls over the behaviour of DNA which develop during early growth seem to be lost. Instead of being specialized to produce just a few types of protein, the rapidly dividing cells start making substances they do not normally manufacture. The best known examples of this phenomenon are tumours of the lung which not uncommonly suddenly start manufacturing hormones which are normally produced only by the specialized cells of the pituitary gland. ACTH and ADH (see index) are frequently manufactured. The ACTH stimulates the adrenal cortex to abnormal activity and so the patient with lung cancer may present with Cushing's syndrome. The ADH prevents normal water excretion and so the patient becomes waterlogged.

The study of nucleic acids and of protein synthesis is thus not simply an academic problem. It contains the seeds of the answers to cancer and to viral disease.

D

7

The liver

As far as biochemistry is concerned, the liver is by far the most important organ in the body. A list of the things which it does, which are mentioned elsewhere in the book, is given in table 7.1. This chapter deals with some of those liver functions which are not dealt with elsewhere.

Table 7.1 *The main biochemical functions of the liver not discussed in chapter 7*

Protein metabolism
1. Deamination of proteins, and manufacture of urea from ammonia.
2. Synthesis of plasma proteins.
3. Synthesis of nonessential amino acids.

Carbohydrate metabolism
1. Synthesis and storage of glycogen and regulation of blood glucose.
2. Oxidation of carbohydrate.
3. Conversion of carbohydrate to fat.

Lipid metabolism
1. Synthesis of fatty acids and phospholipids.
2. Oxidation of fats.

Nucleic acid metabolism
1. Synthesis of the bases found in the nucleic acids.

Structure

The liver consists of columns of cells all of which appear to be identical in structure and function. Serving these cells by carrying to them or from them fluids of various sorts are five separate systems of tubes.

1. PORTAL VENOUS SYSTEM. The portal vein collects blood from the gut which after a meal is rich in digested food. On entering the liver, the portal vein breaks up into smaller and smaller vessels eventually reaching flabby, thin-walled structures known as sinusoids.

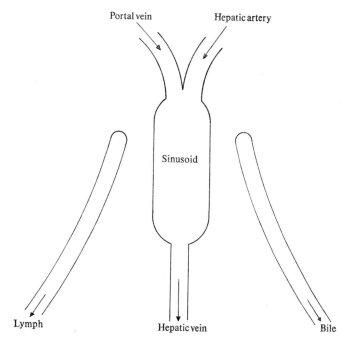

Fig. 7.1. The various systems of vessels which serve the cells of the liver.

These allow the blood to come into contact with the liver cells. The sinusoids contain the so-called reticulo-endothelial cells in their walls. The reticulo-endothelial cells are phagocytic: that is to say that they can engulf solid particles from the blood. Similar cells, part of the reticulo-endothelial system, are found in sinusoids in the bone marrow and spleen. The reticulo-endothelial cells in the liver are particularly important in the removal of old red blood cells and of bacteria which sometimes enter the portal blood from the gut.

2. HEPATIC ARTERIES. The total flow of blood through the liver after a meal is in the region of 1500 ml/minute. About four-fifths of

this comes from the portal vein and is partly deoxygenated as it has already passed through the gut. In contrast, the hepatic arteries supply about 300 ml/minute of freshly oxygenated blood direct from the aorta. The hepatic arteries too send their blood to the liver sinusoids.

3. HEPATIC VEINS. These collect the blood from the sinusoids and pour it into the inferior vena cava which returns it to the right side of the heart.

4. BILIARY TRACT. This is a blind-ended system of tubes which receives the substances secreted by the liver cells. It starts as very tiny tubes in close contact with the liver cells. These tubes join together eventually, forming the important ducts shown in fig. 7.2.

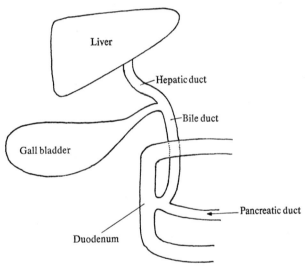

Fig. 7.2. The bile system which carries bile from the liver to the duodenum.

The bile is emptied into the duodenum. The gall bladder is a blind sac whose main function is bile storage. It does not secrete bile itself but it does remove some of the water from bile so making it more concentrated.

5. LYMPHATICS. These collect any excess intercellular fluid and return it to the blood.

Composition of bile

Bile is a substance whose importance lies partly in its role in the digestion of fats and partly in the fact that it is the route by which many substances are excreted from the body. The bile salts (sodium salts of glycocholic and taurocholic acids) act as detergents and help in the digestion of fats (chapter 4). The important substances excreted in the bile are:

1. THE BILE PIGMENTS. These are mainly formed by the breakdown of the haem part of the blood pigment, haemoglobin. The brownish-yellow bilirubin is formed first. When stored in the gall bladder

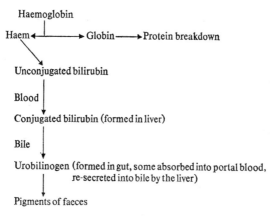

Fig. 7.3. The metabolism of the bile pigments.

bilirubin may be converted to the dark green biliverdin. The bile pigments are responsible for the normal brown colour of the faeces and will be further discussed later in the chapter.

2. CHOLESTEROL. This steroid is secreted in large amounts by the bile.

3. ALKALINE PHOSPHATASE. This is an enzyme which is made in the bones and which is removed from the blood by the liver. If the liver is damaged or if bile excretion is obstructed, the concentration of alkaline phosphatase in the blood rises.

4. PENICILLIN AND AMPICILLIN. Because it is excreted in the bile, ampicillin is often used for biliary tract infections.

5. VARIOUS IODINE-CONTAINING COMPOUNDS. These are artificially manufactured substances which are opaque to X-rays: this
means that they can be identified in the body by the shadows they
throw on X-ray films. Some of the substances used are very opaque
indeed and when injected intravenously they outline first the biliary
tree and then the gall bladder. Others, usually given by mouth are
not so opaque. They cannot be 'seen' in ordinary bile when it is first
secreted and so they cannot be used to outline the biliary tree.
However, when the bile is concentrated in the gall bladder they can
be identified and so can be used to outline the bladder and to assess
its ability to concentrate the bile.

The gall bladder

This does not secrete bile. It merely stores it and removes some
water from it so making it more concentrated. Concentrated solutions tend to deposit substances dissolved in them in the form of
solid material. In the gall bladder such solid deposits make up the
familiar gall stones. The major constituents of a gall stone are
cholesterol and the bile pigments. These substances allow X-rays to
pass through them and so they cannot be seen on a plain X-ray.
Only the 10–15 per cent of gall stones which contain moderate
amounts of calcium can be seen in this way. The stones which are
not radio-opaque can be seen by giving one of the iodine compounds
concentrated in the gall bladder. These normally fill the bladder
producing a smooth, solid shadow. If many gall stones are present,
the bladder will not be filled evenly and the shadow will appear to
have many 'holes' in it.

The factors which make some people form many gall stones are
as yet poorly understood. They tend to be commonest in fat, middle-
aged ladies who have borne many children, but no one can be said
to be exempt. Infections of the biliary tract seem in some cases to
be associated with gall stones.

Other functions of the liver

The most important functions of the liver not yet mentioned are:

1. STORAGE. The liver stores glycogen, iron and vitamins A, D,
and B_{12}.

2. RED CELL DESTRUCTION. The reticulo-endothelial cells in the sinusoids are an important part of the system which removes old red cells from the blood and destroys them.

3. RED CELL MANUFACTURE. During foetal development before birth, the liver is an important site of red cell manufacture. By the time of birth, however, this function has been completely taken over by the bone marrow. Rarely, if the bone marrow is destroyed, the liver may take up its old role and start manufacturing red cells again.

4. DESTRUCTION OF HORMONES AND DRUGS. Many drugs and hormones are inactivated in the liver. Some are broken down completely while others are combined with another substance (such as glucuronic acid) which inactivates them. It is therefore important to be particularly careful when giving drugs to patients with liver disease. A normal dose for a normal person may be an overdose for such a patient because he cannot destroy the drug properly.

5. MANUFACTURE OF CLOTTING FACTORS. The liver is the place where most of the clotting factors essential for blood coagulation are manufactured.

6. MANUFACTURE OF PLASMA PROTEINS. Many plasma proteins and in particular the plasma albumin are made in the liver.

Jaundice

Red cells are destroyed by the reticulo-endothelial cells of the body, particularly in the liver, spleen and bone marrow. The haem part of the haemoglobin is converted into bilirubin which itself is insoluble in water. The bilirubin which is not made in the liver itself is therefore carried around to the liver in combination with the plasma proteins. Bilirubin in combination with plasma protein cannot be excreted by the kidneys and so none appears in the urine. This type of bilirubin is sometimes known as 'indirect' bilirubin because of its behaviour in the van den Bergh test for bilirubin in blood.

In the liver the bilirubin is extracted from the blood and combined (conjugated) with glucuronic acid to give bilirubin glucuronide. Bilirubin glucuronide is often known as conjugated or 'direct' bilirubin. Conjugated bilirubin is soluble in water and is excreted in the bile. Normally it does not enter the blood and so does not reach the urine. If it did gain entry to the blood, conjugated bilirubin could be excreted in the urine as it is freely soluble in water.

In the intestine, conjugated bilirubin is converted to a mixture of substances collectively known as urobilinogen. These are soluble in water and may be absorbed by the portal vein. However, the urobilinogen is normally promptly removed from the portal blood by the liver and re-excreted into the bile. Normally urobilinogen does not enter into the general circulation nor, even though it is soluble in water, does it enter the urine.

Jaundice is the term given to the yellow pigmentation of the skin, white parts of the eyes and mucous membranes which occurs when the plasma contains unusually large amounts of either unconjugated or conjugated bilirubin. There are three main types of jaundice:

1. HAEMOLYTIC. This occurs when the red cells are being destroyed at an unusually rapid rate. Liver function is normal and the bilirubin in the blood is in the normal unconjugated form and is bound to plasma protein. No bilirubin therefore appears in the urine. No urobilinogen appears in the main circulation or in the urine except very occasionally when the rate of red cell destruction is so high that the liver cannot remove all the urobilinogen it receives in the portal blood. The faeces are usually very dark because of the excess bile pigment.

2. HEPATOGENOUS. This occurs when the liver cells are damaged. There are many causes, the commonest being viral hepatitis, cancer and cirrhosis (destruction of the liver, sometimes due to excessive alcohol intake, but often of unknown cause). The flow of substances from the liver cells into the bile is disrupted and so some conjugated bilirubin, normally strictly confined to the bile, manages to enter the blood. Since conjugated bilirubin is soluble in water, it also appears in the urine. Furthermore, urobilinogen cannot be completely extracted from the portal-vein blood by the damaged liver and so it too appears in the general circulation and in the urine. This may not be true if the liver damage is very severe: then very little bile pigment reaches the gut at all and virtually no urobilinogen is formed so none can appear in the urine. Because of the lack of bile pigment the faeces are pale.

3. OBSTRUCTIVE. This is the result of mechanical obstruction of some part of the biliary tract, most often because of a gall stone but

sometimes due to a tumour (e.g. of the pancreas). Some of the dammed up bile forces its way into the blood and so conjugated bilirubin again appears in the blood and the urine. There is not usually any urobilinogen in the urine: none is formed as no bile pigments can enter the gut because of the blockage. The faeces are usually very pale ('clay-coloured').

8

The Vitamins

During the nineteenth century it became clear that diets containing more than adequate amounts of highly purified fats, proteins and carbohydrates could not maintain health. Animals fed such diets developed a whole variety of diseases. Yet these diseases could be cured by the addition of minute quantities of substances known as vitamins to the food. Vitamins are organic compounds which are not oxidized by the body to supply energy, nor are they part of the structure of the body. They cannot be manufactured by the body in sufficient quantities and so they must be supplied in the diet. They are essential for normal metabolism and many of them, particularly those of the B group, act as coenzymes. This means that they work closely with an enzyme: they may bring to it atoms or small groups of atoms required for addition to a substrate or they may carry away similar atoms or small groups which may be split off from the substrate by the enzyme.

Many of the reactions for which vitamins are required are common to most of the cells in the body. An established cell which already has its full quota of vitamins will not initially suffer unduly if the supply of the vitamins to the body is reduced. Only if the deficient intake of the vitamin continues for a long time will established cells show any effects as the vitamin molecules within them are gradually destroyed in the process of wear and tear. In contrast to the established cells, when the intake of a vitamin falls, those cells which are being newly manufactured will have inadequate amounts and will soon show the effects of the deficiency. Some tissues of the body are always manufacturing new cells to replace old ones which have been destroyed. The surfaces of the skin and the lining of the gut are quite quickly worn away and the cells there must be continually replaced by the formation of new cells. Similarly the red and many of the white cells of the blood are also continually being destroyed and must continually be replaced. Because of this most vitamin deficiencies first show themselves in the damage they do to the skin, the lining of the gut and the bone marrow where the red cells are manufactured. Virtually all vitamin deficiencies are therefore characterized by anaemia and by skin diseases.

When vitamins were first discovered, their chemical composition was unknown and so they were given letters, A, B, C, D and K. We now know the chemical composition of most of the vitamins and the letters are beginning to be replaced by the proper chemical names. However, it will be many years before the much simpler letters fall out of use.

The vitamins are usually divided into two great groups, those which readily dissolve in fats (A, D, E and K) and those which readily dissolve in water (B and C). The water-soluble ones are usually common in meat, cereals and vegetables while the fat-soluble ones occur primarily in dairy produce and fish-liver oils. The fat-soluble ones are absorbed from the gut in combination with fats and so anything which leads to a failure of fat absorption (e.g. lack of bile or pancreatic juice) leads to a deficiency of vitamins A, D, E and K.

THE B COMPLEX

The substance originally given the name of vitamin B is now known to be a mixture of many different chemically distinct compounds. However, these compounds often occur together in foods and when one is deficient the others are usually deficient too. Disease due to a pure deficiency of a single B vitamin is rare. The B-complex vitamins will therefore be briefly described, separately at first, and then the diseases which occur in B-group deficiency will be discussed at the end of the section. The common sources of each vitamin are listed in table 8.1 which is shown on the next page.

Thiamine (aneurin, vitamin B₁)

This is essential for many biochemical reactions but in particular for the conversion of pyruvic acid to acetyl coenzyme A (chapter 3). In thiamine deficiency, therefore, it is difficult to oxidize carbohydrate properly. Since the nervous system depends entirely on carbohydrate for its energy supply, it is not surprising that it ceases

Table 8.1 *The main sources of vitamins in the food*

A	Dairy produce, fish-liver oils.
D	Dairy produce, fish-liver oils.
K	Widely distributed in small amounts. Made by gut bacteria.
E	Widely distributed, particularly vegetable oils and green vegetables.
Thiamine	Wholemeal flour, cereals, peas, beans, yeast.
Nicotinamide	Liver, meat, wholemeal flour.
Riboflavin	Meat, milk, wholemeal flour.
Pyridoxine	Very widespread.
Pantothenic acid	Liver, eggs, meat, milk.
Biotin	Widespread.
Folic acid	Widespread in small amounts. Made by gut bacteria.
B_{12}	Liver and all foods of animal origin.
C	Fresh fruits and vegetables.

to function properly in thiamine deficiency. The heart also depends heavily on thiamine and so there are defects in cardiac function. Thiamine is not destroyed by ordinary cooking but pressure cooking and canning remove much of it from the food.

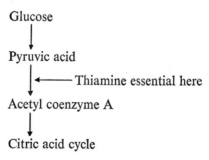

Fig. 8.1. One of the roles of thiamine in metabolism.

Nicotinamide (niacin)

This too is essential for carbohydrate oxidation and for the operation of the citric acid cycle. It is very resistant to cooking. It is only a partial vitamin in the sense that the body can manufacture part of its daily requirement from the amino acid tryptophan. Only if the diet is deficient both in nicotinamide and tryptophan will signs of deficiency appear.

Riboflavin

This also is required for oxidation of foodstuffs. With most diets an important part of the supply comes from milk. It is moderately resistant to cooking but is rapidly destroyed by exposure to bright light. Ninety per cent of the riboflavin in a bottle of milk may be destroyed if the bottle stands in the sun for a morning. In man another important source of riboflavin is the bacterial colony in the intestine which manufactures it. Sterilization of the gut by anti-biotics may greatly reduce the riboflavin intake.

Pyridoxine (vitamin B_6)

This is important for the manufacture and interconversion of amino acids and also for the utilization of the essential fatty acids. It is very widespread in ordinary foods but some tropical diets and some synthetic infant diets may be deficient in it. Isoniazid, a drug important in the treatment of tuberculosis, can inactivate pyridoxine. Signs of vitamin deficiency may therefore appear even though the intake in the food is normal.

Pantothenic acid, lipoic acid and biotin

Most diets contain adequate amounts of these vitamins and so deficiencies are highly unlikely. Biotin cannot be absorbed from the gut in the presence of raw egg white. The only recorded case of deficiency occurred in a man whose diet consisted entirely of a dozen raw eggs and several pints of wine daily!

Folic acid

This is essential for the manufacture of nucleic acids and so is particularly required when cell division and protein synthesis are occurring rapidly. It is widely distributed in the food but much is made in the gut itself by the resident bacterial colony. It is particularly required by the skin and the lining of the gut, by the marrow manufacturing red cells and by the pregnant mother acting as host for the growth of a new child. Deficiencies are most likely to occur in infancy, before the normal bacterial colony has been established, and in pregnancy when requirements are very high. A deficiency results in a type of anaemia in which not enough red cells can be manufactured. Haemoglobin synthesis is less affected and so there is too much haemoglobin for the number of cells being formed: the cells therefore tend to be larger than normal. This type of anaemia is known as megaloblastic. It is also not uncommon in epileptics: many of the drugs used in epilepsy interfere with action of folic acid. In the tropics there may be changes in the gut bacteria which leads to the colony using more folic acid and manufacturing less. The deficiency damages the lining of the gut leading to a condition known as tropical sprue, one of whose symptoms is persistent diarrhoea.

Rapidly dividing cancer cells also need large amounts of folic acid. Drugs which interfere with the action of folic acid are therefore sometimes used as anti-cancer agents.

Vitamin B_{12} (cyanocobalamin)

This is an unusual vitamin in that it is very widely distributed and is required in such minute amounts that natural diets are never deficient in it. The only people likely to suffer from a dietary deficiency are those extreme vegetarians known as vegands who do not eat any animal products (including milk and eggs). In spite of this, disease due to vitamin B_{12} deficiency is very common. This paradoxical situation arises because a special absorptive mechanism is required to get the vitamin across the gut wall into the blood. The stomach secretes a substance known as intrinsic factor which combines with the vitamin B_{12} in the food. The combination of intrinsic factor and vitamin is then bound to the wall of the small intestine and the vitamin is taken across into the blood. In the absence of

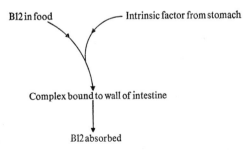

Fig. 8.2. The absorption of vitamin B_{12} from the gut.

intrinsic factor the vitamin cannot become bound to the wall of the intestine and so is hardly absorbed at all even though there may be plenty in the diet. A failure of intrinsic factor secretion is not uncommon and gives rise to the disease known as pernicious or Addisonian anaemia.

Vitamin B_{12}, like folic acid, is required for cell division and so the anaemia is of the megaloblastic type. Megaloblastic anaemia may therefore be caused by a deficiency either of folic acid or of B_{12}. Pernicious anaemia is caused only by B_{12} lack: in addition to the anaemia, peripheral nerves and parts of the spinal cord degenerate. Vitamin B_{12}, unlike folic acid, seems to be essential for the functioning of these parts of the nervous system.

B-complex deficiencies

There are two major diseases which are caused by B-complex deficiency. Beri beri is primarily due to thiamine deficiency but is always aggravated by lack of other B vitamins. Pellagra occurs primarily where corn (maize) is the staple diet and is thought to be due to a combined absence of nicotinamide and tryptophan although again the lack of other B-complex members almost certainly plays a part.

Beri beri itself occurs in two main forms. In the 'dry' variety, the main features are due to defects in the nervous system. Muscles become wasted and paralysed and sensation is lost. The brain may also be damaged, resulting in lethargy, loss of appetite and memory loss. In developed countries, beri beri is likely to occur only in chronic alcoholics who gain all their calories from alcohol and who eat no proper food. In the 'wet' form of beri beri, the main organ to suffer is the heart. This cannot pump blood around the body effectively. As a result the kidneys fail to excrete fluid normally and large amounts of water accumulate in the form of oedema.

With pellagra, the early signs of the disease are seen in the skin and in the lining of the gut. The mouth and tongue become red and ulcerated and there is vomiting and profuse diarrhoea. Muscles become weak and sensation is affected. Finally the brain is damaged and the patient becomes frankly mad. As with beri beri the end result of the untreated disease is death.

VITAMIN C (ASCORBIC ACID)

Scurvy is a disease which has been well known for hundreds of years.

It has always tended to occur in people deprived of fresh food for long periods such as those going for long sea voyages and those in northern cities in winter. Today in the developed countries the disease may sometimes be seen in lonely old people who eat everything out of tins. Scurvy is due to a deficiency of vitamin C. The vitamin is found primarily in fresh fruit and vegetables. Since it is very easily destroyed by cooking or canning, cooked food may contain very little of it. Freezing, in contrast, has little effect on the vitamin C content of food.

The main effects of ascorbic acid deficiency can be accounted for by the fact that it is essential for the manufacture of the structural protein, collagen. Collagen is important in bones and joints, in tendons and ligaments, in the support of blood vessels and of the skin and in the healing of wounds (white scar tissue consists largely of collagen). It is therefore not surprising that the main features of scurvy are bleeding around the gums and into the skin, painful joints and a failure of wounds to heal quickly. In children the skeletal system fails to develop normally because of the lack of collagen for bone. In addition to its role in collagen synthesis, vitamin C appears to be essential for the normal function of folic acid. Thus anaemia is also a common feature of scurvy.

VITAMIN A

Vitamin A is found in large amounts in dairy produce and in fish-liver oils. In addition, all green or yellow plants and vegetables contain substances called carotenes which are not active vitamins themselves but which can be converted to vitamin A in the body. The vitamin is stored in large amounts in the liver. The main consequences of vitamin deficiency are as follows:

1. Night blindness develops. Vitamin A is essential for the normal functioning of the rods, the elements of the retina at the back of the eye which are required for vision in dim light. The vitamin is not required for the functioning of the cones with which we see when the light is bright. The earliest complaint of the patient is often therefore of inability to see in the dark.

2. The vitamin is essential for the normal moulding of bone in growth. In its absence the skull, and the holes in it which allow the nerves to pass out and in, fail to enlarge normally. The brain and the nerves therefore become damaged.

3. The skin and the mucous membranes lining most of the tubes in the body (e.g. the gut and the bronchi of the lungs) become thick, dry and liable to infection. The cornea of the eye becomes cornified and blindness may occur. Digestion is poor and because of damage to the respiratory tract, pneumonia is a common cause of death.

4. Males are infertile and while females can conceive they cannot carry normal infants to term.

In these days of readily available enriched sources of vitamins, it is not impossible, especially in infants, for an overdose to occur. The characteristics are an itchy rash, severe headache, weakness, loss of appetite and pain in the long bones. The condition is reversed within a few days of stopping the high intake of the vitamin.

VITAMIN D

Vitamin D_2 (calciferol) and vitamin D_3 are closely related compounds, both of which have vitamin activity. There is also another substance, 7-dehydrocholesterol, which is common in dairy foods and fish-liver oils. In itself it is inactive but it is converted to vitamin

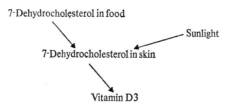

Fig. 8.3. The manufacture of vitamin D in the skin.

D_3 on exposure to ultraviolet light: the conversion can readily occur in skin exposed to sunlight.

Vitamin D is important to the body in two main ways:

1. It is essential for the absorption of calcium from the gut. In its absence sufficient calcium cannot be absorbed to replace even the small calcium losses which occur into the urine and faeces in adults. In pregnancy, and in infancy and childhood when calcium requirements are high for lactation and new bone formation, the lack of calcium is particularly important.

E

2. It is essential for the manufacture of normal new bone and for the incorporation of calcium into bone.

Two main types of disease can result from a lack of vitamin D, rickets in children and osteomalacia in adults. Rickets is a disease in which the growing points (epiphyses) of bones do not function properly and as a result the bones become twisted. This leads to permanent deformities which persist into adult life. In adults whose epiphyses are no longer active, rickets cannot occur. However, the persistent lack of calcium leads to softening of the bones and particularly of the pelvis which has to bear so much of the weight of the body. The deformed pelvis may narrow the outlet for the foetus and cause difficulties in childbirth.

When rickets was first being studied, there was considerable argument about its cause. One group maintained that lack of sunlight in narrow, smoke-blanketed city alleys was the cause. The other said that rickets was due to a poor diet. In fact it is now apparent that both ideas were correct. If there is enough D_2 and D_3 in the diet, rickets will not occur. On the other hand, 7-dehydrocholesterol is found in the skin: if the skin is exposed to ample sunlight, again enough D will be formed and rickets will not occur.

Vitamin D is perhaps the vitamin which is most likely to be deficient in what seems to be a normal diet. This is especially true in infancy and in pregnant and lactating women. Margarine is therefore artificially enriched with the vitamin in most countries and it is usually a routine precaution to give extra amounts of the vitamin to infants and to nursing and pregnant mothers.

VITAMIN K

Vitamin K is a blanket expression which covers a number of substances. Most of these are naturally occurring but one, K_3 or menadione, is artificially synthesized. Vitamin K is widely distributed in food and quite large amounts are manufactured by the bacteria in the gut. The only situations when K deficiency is likely are:

1. In the first days of life when there are no bacteria in the gut. Infants in whom there is any risk of bleeding (e.g. due to birth trauma) are routinely given an injection of vitamin K.

2. Following sterilization of the bowel by antibiotics.

3. In obstructive jaundice or in any other condition when the absorption of fatty substances as a whole is defective.

4. In the presence of drugs which interfere with the action of vitamin K. These occur naturally in some plants and cows have become K deficient after eating these. Drugs based on the substances isolated from these plants are now deliberately used in medicine to interfere with blood coagulation in situations in which there is an abnormal tendency for clotting to occur.

The main action of the vitamin appears to be in the synthesis of clotting factors by the liver (particularly prothrombin and factors VIII, IX and X). In the absence of these factors, clotting is defective and abnormal bleeding is likely to occur.

VITAMIN E

This too is a blanket expression for a group of compounds known as the tocopherols. They are very widely distributed and a natural deficiency is most unlikely except possibly in those who over a very long period are incapable of absorbing fat. The main effects of deficiency which have been studied in animals and man are infertility in both sexes, muscle weakness and anaemia. There is no evidence that human infertility is ever due to vitamin E deficiency.

9

Nutrition

In designing a good diet, at least four major problems must be dealt with.

1. Sufficient water must be provided daily. No one ever thinks of this as a nutritional problem yet it should be remembered that a lack of water kills in days, whereas a lack of food usually takes weeks to have this effect.

2. Foodstuffs which can be oxidized to provide energy must be supplied in adequate amounts. The energy requirement is usually expressed in terms of kilocalories. Most of the energy comes from the oxidation of fat and carbohydrate with much smaller amounts coming from protein.

3. A diet which supplied enough calories may still not contain all the substances required for growth in children and the maintenance of full health in adults. The essential amino acids, essential fatty acids, minerals and vitamins must also be present.

4. A diet which completely fulfils the above three conditions may nevertheless still be defective. Instead of containing too little of a substance it may contain too much. Natural substances which may cause disease when taken in excess include fat, carbohydrate and vitamins A and D. But increasingly the food contains unnatural substances which should not be there. Some of these, such as the insecticides, get there purely by accident: they are so widely distributed on the earth's surface that it is impossible to stop them getting into the food. Others such as cyclamates and saccharin may be put in as sweeteners, preservatives or colouring matter. In most cases we do not know whether or not these substances are harmful.

The second and third problems face the less developed areas of the world. The fourth is primarily a problem met with in the developed countries.

Calorific requirement and metabolic rate

The numbers of calories released by pure samples of the various types of food can be measured by burning a known weight of the food in a device known as a bomb calorimeter. It has been demonstrated that when food is burned in such a device it releases precisely

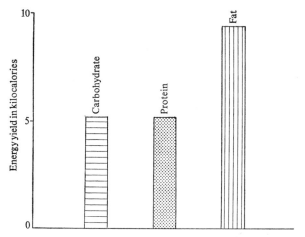

Fig. 9.1. The relative amounts of energy which are freed by the burning of 1 g of fat, 1 g of carbohydrate and 1 g of protein.

the same number of calories as does the same amount of food when oxidized in the body itself. Each gram of protein burnt yields about 5·3 kcal. Fat yields about 9·3 kcal/g while carbohydrate yields about 5·3 kcal/g. Thus a given weight of fat releases more calories than a given weight of any other food and it is therefore not surprising that excess food is stored in the form of fat.

There are a number of ways of measuring the number of kilocalories worth of food which the body consumes. The simplest one depends on the fact that for every litre of oxygen used up by the body, about 4·8 kcal are produced. The person under test is asked to breathe from a bag or a cylinder containing a known amount of oxygen. After a measured time, the amount of oxygen left is measured and so the amount of oxygen used up can be easily calculated.

The number of kilocalories produced in that time can then be worked out by multiplying the number of litres of oxygen used by 4·8.

The metabolic rate of a person is the number of calories he or she uses up per day when carrying out normal activity. The number of calories taken in in the food can be estimated by a careful analysis of the diet: the weight of fat in grams is measured and multiplied by 9·3 to give the number of kilocalories which could be produced by the oxidation of that amount of fat: the number of kilocalories which could be released by carbohydrate is estimated by measuring the weight of carbohydrate in grams and multiplying it by 5·3: the potential energy yield of the protein is estimated in the same way. Thus the amount of calories which could be released by the amount of food taken in can be compared with the number of calories actually used by the body each day. We can represent the balance between the two by means of the following equation:

Calorific value = Number of calories + Calorific value of
of food   used by the body   fat stored

This means that if the food contains more calories than are used up in a day by bodily activity, the extra calories are stored away as fat. If the food contains too few calories to supply the energy needs for that day then some of the fat stores are broken down to make up the difference. It is impossible to avoid the consequences of this equation. There is no easy road to the loss of weight. The only way to lose weight is to reduce the food intake below the level of energy expenditure. This can be done either by taking the same amount of exercise but eating less or by eating the same amount but carrying out more exercise, or of course by combining the two things. It is as simple (or as difficult!) as that. Those who are fat eat too much in relation to the amount of exercise they do. Those who are put on a diet and still do not lose weight are either still eating too much or not exercising enough. Weight will always be lost if calorie expenditure rises above calorie intake. No tricks can be used to get around this fundamental relationship.

Factors which alter metabolic rate

As mentioned earlier, the metabolic rate of a person is the number of calories used up per day. The basal metabolic rate is something rather different which attempts to express something more fundamental about the metabolism of the body. For example, the crude metabolic rate of a large healthy man might be in the region of 3,000 kcal/day while that of a healthy child might be 1,000 kcal/day. That simply means that more calories are required to run a large body and does not imply that there is something fundamentally different between the metabolism of a man and that of a child.

In trying to estimate the basal metabolic rate (BMR), the first thing is to eliminate differences which occur because of differences in muscular activity and food intake, by keeping the person at rest and without food for at least 12 hours. This will bring the calorie expenditure of the lumberjack and the businessman down to the same level but still they will have much larger crude metabolic rates than a child. It has been found, however, that the BMR is closely related to the surface area of the body. The number of kilocalories used up in a resting individual per square metre of body surface is remarkably constant for all healthy humans and indeed for all healthy mammals. The BMR is therefore expressed as the number of kilocalories produced per square metre of body surface per day. The normal figure is close to 1,000 kcal/m²/day. It can easily be measured indirectly by measuring oxygen consumption.

The following factors may alter the BMR:

1. It falls very slowly with increasing age.

2. It is very slightly lower in women than in men.

3. It may increase on prolonged exposure to cold weather. The body thus produces more heat for the maintenance of its normal temperature.

4. It rises by about 10 per cent for every 1 °C rise in body temperature. The BMR therefore may be greatly elevated in fever.

5. It is affected by the level of thyroid hormone. When the output of thyroid hormone is too high (thyrotoxicosis) the BMR may be 50 per cent or more above normal. When the hormone output is too low (myxoedema), the BMR may be well below normal. BMR measurements used to be made during the diagnosis of thyroid disease but they have now largely been superseded by other ways of measuring thyroid activity.

In normal individuals, most of the above factors which can alter metabolic rate are overshadowed by the much larger changes caused by food intake and exercise which by definition are excluded from the BMR. Even if an individual remains completely at rest, taking in food increases his metabolic rate. This is known as the specific dynamic action of food and it seems to occur because energy is required for the digestion, absorption and processing of food. The energy used up in this way in making the food available for use by the body is clearly not available for other bodily activities. Thus only about 90 per cent of the calorific value of food can actually be used by the body. Even the rise in metabolic rate on taking in food is totally overshadowed by the extra calories required for even mild exercise. Work at an office desk requires an extra 40 kcal/hour above basal levels while the shovelling of earth might use up an extra 300 kcal/hour or even more. Therefore in calculating the total calorific requirement of an individual the following factors must be taken into account:

1. BMR required to keep the body ticking over. For an average sized man this might be 1700 kcal/day.

2. Extra calories required because of specific dynamic action of food. An extra 10–15 per cent over the basal level would be required, say an extra 200 kcal, making 1900 kcal/day.

3. Extra calories required for muscular activity. Assume that the person spends 8 hours sleeping, 8 hours in leisure activity and 8 hours working.

    a. Sleep requires no extra calories.

    b. Leisure energy expenditure clearly varies widely, but a reasonable estimate might be an extra 50 kcal/hour. Eight hours leisure would therefore add 400 kcal making 2,300 in all.

    c. Eight hours work. An office job might add $8 \times 40 = 320$ kcal while lumberjacking might add $8 \times 300 = 2,400$ kcal. This therefore produces the really big calorific requirements which range from about 2,600 kcal/day for a light office job to about 4,700 kcal/day for heavy labour.

4. Finally 10–15 per cent of the calorific value of unprepared food is lost, being destroyed by cooking, left on the side of the plate or passing straight through the gut to the faeces. Therefore the calorific value of the unprepared food must be 10–15 per cent higher than the calorific requirement of the body.

Supplying calories

Approximate daily calorific requirements of children and of adults doing various sorts of work are shown in table 9.1. In developed countries about 10–15 per cent of the calories are supplied by protein. The remainder are supplied by fat and carbohydrate, the relative proportions of the two depending to some extent on the economic status of the individual. Carbohydrate-rich foods tend to be much cheaper than fats. Many poorer people therefore obtain almost all their calories from such foods as bread and potatoes.

Table 9.1 *Some approximate daily kilocalorie requirements for children and adults*

Children, 1–2 years	1,000
Children, 2–3 years	1,250
Children, 3–6 years	1,550
Children, 6–8 years	1,850
Children, 8–10 years	2,150
Children, 10–12 years	2,550
Children, 12–14 years	2,900
Girls, 14–18 years	2,900
Boys, 14–18 years	3,200
Housewives	2,700
Nurses, men doing light work	3,000
Men doing heavy work	4,000

With increasing income, fat-rich foods such as dairy products and meat, can be bought in larger amounts and as much as 40–50 per cent of the total calorie requirement may be supplied in the form of fat. In most developed countries, the problem is that most people take in too many calories for the amount of exercise that they do and hence tend to become obese. In less fortunate areas of the world, a low calorie intake is much more significant.

Starvation

Most of the people in the poorer parts of the world are malnourished rather than completely starving. They receive neither enough calories nor enough of essential dietary constituents such as amino acids, vitamins and minerals. A number of experiments have revealed the devastating effects in adults of a simple reduction in calorie intake to the level of 1,500 kcal/day. The human volunteers in these experiments were previously completely healthy and their diet contained more than enough protein, vitamins and minerals. The only defect was a lack of calories. Quite soon these previously fit men and women became weak, lost weight, and were incapable of making any sustained mental or physical effort. It is frightening to think of the consequences when a whole nation is in this state of health.

Although undernourishment is much commoner than total starvation, the latter is by no means unknown. If the intake of water is normally maintained, a normal lean individual will die within 4–6 weeks if he is totally cut off from food. The stages in the process are as follows:

1. Glycogen stores are used up within the first couple of days.

2. The body must then rely on fat oxidation for its energy supply. There may not be enough oxaloacetic acid for the entry of acetyl coenzyme A into the citric acid cycle. The ketone bodies (chapters 4 and 13) then tend to accumulate and ketosis may occur.

3. Protein is broken down only very slowly but its degradation does occur and the muscles which are largely protein gradually lose their bulk. When the fat deposits have been finally used up, protein is the only food left. It must therefore be broken down much more rapidly and there is a sharp increase in the rate of production and excretion of urea. This is known as the 'pre-mortal rise' because it indicates that death cannot be far off.

One of the outstanding features of starvation is the accumulation of oedema fluid in the legs and abdomen. The bloated abdomen looks particularly tragic against the background of the remainder of the emaciated body. The cause of the oedema is uncertain. Part may be due to the weakness of the heart which prevents the normal excretion of fluid by the kidneys. Part is due to the fall in the plasma protein concentration which occurs during starvation. The osmotic pressure of the blood thus falls and this allows fluid to escape into the tissues.

Kwashiorkor

No human disease is known to occur because of a deficiency of a single amino acid. However, a generally low protein intake is of major importance in the disease kwashiorkor which is not uncommon in South America, Africa and Asia. The word is said to mean 'displaced child' for the disease usually develops when a child is prevented from breast feeding by the arrival of a younger infant. The outstanding characteristics of the disease are retardation of growth, anaemia, fatty degeneration of the liver and a deficiency of pancreatic enzymes. In Negro children the hair is usually reddish instead of its normal black colour. Most of the changes are obviously related to protein deficiency. The lack of digestive juices obviously clearly makes things worse for it is impossible to digest and absorb properly even the small amounts of protein which may be taken in with the food. In adults protein deficiency leads to muscle wasting, liver degeneration, oedema and a hopeless feeling of lassitude.

ESSENTIAL DIETARY CONSTITUENTS

Apart from the necessity of providing enough calories, all diets must contain certain additional materials which the body cannot manufacture for itself. The main ones will be briefly discussed in this section.

Essential amino acids

There are ten amino acids which must be provided in the diet because they cannot be manufactured by the body in sufficient quantities to meet its needs. Proteins which contain all the essential amino acids are said to be of high biological value. Those which do not contain them all are said to be of low biological value. In general, proteins of animal origin such as those of meat and milk are of high biological value. Those of vegetable origin tend to be of low biological value. However, two different vegetable proteins are not usually lacking in the same amino acids and they can therefore make up one another's deficiencies. Thus, two vegetable proteins of low biological value may together make up a meal of high biological value. It is important to realise that all the essential amino acids must be present in the same meal if they are to be used for protein synthesis. Amino acids which cannot be used at once are promptly broken down by the liver and wasted. Therefore it is pointless to give a protein of low biological value at one time and then another different low-biological value protein 3 hours later. The amino acids from the first one will be broken down before the second arrives on the scene.

Elements

The main elements which must be provided in the food are sodium, potassium, chlorine, calcium, phosphorus, iron, iodine, magnesium, copper, manganese and fluorine. Of these, calcium, iron, iodine, magnesium and fluorine are particularly important because deficiencies can occur in practice as well as in theory. Deficiencies of the other substances are extremely unlikely to occur naturally. The main situations in which deficiencies of elements may occur are as follows:

1. Iron is essential for the manufacture of the haemoglobin in the red cells of the blood. It is also required in much smaller amounts

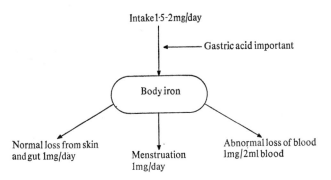

Fig. 9.2. The intake and loss of iron.

for all other cells. As cells are repeatedly being lost from the surfaces of the skin and gut, about 1 mg of iron/day is lost from the body in this way. The average western type diet contains 15–20 mg of iron/ day but absorption from the gut is poor and only about 1·5–2 mg of iron enter the body. The absorption of even this small amount seems to depend on the presence of normal amounts of acid in the stomach. People with chronic inflammation of the stomach (chronic gastritis), those with gastric ulcers, and those who have had part of their stomach surgically removed produce less acid than normal. They cannot therefore absorb iron normally and are likely to develop an iron deficiency anaemia unless extra iron is provided in the diet. Because of loss of iron in the blood during menstruation, all women of reproductive age are also likely to develop a similar anaemia. With most women the amount of blood loss when averaged out over the whole month means an extra loss of about 1 mg of iron/day. Women therefore lose about 2 mg of iron/day in all and are in a precarious state of iron balance.

2. Calcium is primarily found in dairy products and most adults in the western countries probably get sufficient. However, children and pregnant and lactating women need more calcium and must take in

extra supplies, usually in the form of milk. It has recently become apparent that old people living alone also tend to take in too little milk and may become calcium deficient. Their bones may weaken more quickly than is usual in old age.

3. Iodine is essential for the manufacture of thyroid hormone. If there is too little iodine in the diet, the person may become thyroid deficient and the gland may grow in a desperate effort to produce more thyroid hormone. This produces the unsightly swelling in the neck known as a goitre. Iodine occurs on this planet primarily in the sea, and on moving away from the sea drinking water and food contain less and less iodine. Goitre therefore tends to occur in mountainous areas far from the sea such as Switzerland, the Himalayas and Derbyshire in England. In many countries, iodine is added to salt to ensure that no one suffers from a goitre of this type.

4. The importance of magnesium has been recognized only recently. It seems to be essential for the normal functioning of the nervous and muscular systems and in its absence there may be spasms, irritability and eventually unconsciousness. Most normal diets contain more than enough magnesium and deficiency is likely only in two abnormal situations. The first is severe diarrhoea when quite large amounts of magnesium may be lost in the faeces. The second is during the prolonged maintenance of a patient on intravenous fluids which do not usually contain magnesium. The condition is rapidly reversed by an injection of a magnesium salt.

5. Fluoride seems to be essential in only one way, for the development of strong teeth which are resistant to decay. Most natural water supplies contain less than the optimum amount of fluoride to achieve this end and so in many parts of the world small amounts of fluoride are added to the drinking water in order to enable the body to resist tooth decay more effectively.

Essential fatty acids and vitamins

These have already been discussed in other chapters.

PRACTICAL NUTRITION

In planning a diet for a patient, four conditions must be fulfilled:
1. The diet must contain enough calories. Normal daily calorie

requirements are shown in table 9.1 above and the calorie contents
of some well known types of food in table 9.2.

Table 9.2 *The kilocalorie con-*
tents of 100 *g* (*very roughly*
4 *oz*) *of various types of food*

White bread	240
Rice	360
Milk	70
Butter	790
Cheese	420
Steak	270
Fish	170
Potatoes	70
Peas	80
Cabbage	10
Orange	30
Apple	50
Sugar	400

2. The diet must contain enough protein of high biological value,
enough vitamins, enough elements and enough essential fatty acids.

3. It must contain above-normal amounts of some substances in
special cases, e.g. pregnant women may need extra iron, calcium
and folic acid.

4. Some patients may be harmed by certain foods and in these cases
a diet which does not contain these foods must be provided.

Some special nutritional problems which occur frequently are
discussed in the remainder of this chapter.

Obesity

Any observant nurse very quickly learns the dangers of being over-
weight. Many of the patients in the wards suffer from obesity and
this may aggravate or even be the cause of their other diseases. The
insurance companies have demonstrated conclusively that the fatter
a patient is, the more likely is he or she to die at an early age. Obesity

plays a contributory role in many conditions, notably heart disease and high blood pressure. No nurse who has been in an operating theatre needs to be reminded of how much more difficult is the surgeon's task when the patient is obese. Post-operative respiratory problems are also much more common in fat people who find it more difficult to breathe deeply and evenly. Arthritis tends to be commoner in the fat as the joints, especially the legs, have to carry so much more weight. There are virtually no advantages in being fat.

Any obese person has been eating too much for the amount of exercise he or she has been doing and the extra calories have been laid down as fat. Very few obese patients have any hormonal abnormalities which cause their fatness and their state is the result either of gluttony or lack of exercise. The aim of the dietary treatment of obesity is therefore to reduce the calorie intake below the patient's calorie needs so that the extra calories can be obtained by burning up the fat deposits. A drastic reduction in calorie intake can make a person feel weak and ill. On the whole it is better for a patient to lose weight slowly but steadily. It is better to acquire a habit of eating more wisely than it is to attempt a crash diet which is unlikely to be of any permanent value. The crash diet may be briefly successful but it usually makes the patient feel ill. Moreover, since it is not continued long enough to bring about any permanent change in eating habits, the weight is usually rapidly regained as soon as the crash period is over. For all these reasons, working people who want to lose weight should probably attempt a 2,000 or even a 2,500 kcal diet. This will enable them to lose weight slowly but consistently without becoming ill; 1,500 or 1,000 calorie diets should be attempted only by those who are not working and can afford to rest.

Pregnancy

The weight gain in a normal pregnancy is not usually much more than about 60 g (roughly 2 oz)/day. It is therefore quite unnecessary to have a great increase in calorie intake. However, some particular essential food factors, notably iron, calcium and folic acid, tend to get used up much more rapidly and it is wise to take supplements of these. Taking extra vitamins is unlikely to do any harm although for most women on a good diet in developed countries there is little evidence that this is necessary.

Diabetes

The main problem in diabetes is that the body cannot cope normally with carbohydrates. The aim of a diabetic diet is therefore to reduce the intake of carbohydrate. Since many diabetics, especially elderly ones, tend to be obese, the diet usually also aims to bring down body weight. The most dangerous carbohydrates are those like sucrose (cane sugar) which need very little digestion or those like glucose which do not need to be digested at all. These substances are very rapidly absorbed from the gut and the sudden flood puts a great strain on the diabetic's limited ability to cope with carbohydrate. An attempt is therefore usually made to exclude cane sugar from the diet as completely as possible. The only carbohydrates permitted (and these usually in small amounts) are ones like starch which are slowly digested and which enter the blood in a steady trickle.

Renal failure

When the kidneys fail to excrete waste products normally, the body has great difficulty in getting rid of potassium and of urea and other products of protein breakdown. The protein breakdown products affect the nervous system and cause weakness and a comatose state: this is often called uraemia because the blood concentration of urea becomes very high. The potassium affects the heart and may interfere with its normal rhythm.

The major sources of potassium in the diet are fruit and meat and so the intake of these must be cut down. The protein intake must be cut down to the minimum compatible with the maintenance of life. This reduces the rate of urea accumulation. The protein must of course be provided in a form that is of high biological value and the patient with renal failure presents a real challenge to the dietician.

Effects of drugs

A few drugs must be considered in diet. The most important are the anti-depressive agents, the monoamine oxidase (MAO) inhibitors. Monoamine oxidase is an enzyme which normally seems to destroy naturally occurring amines in the body (such as adrenaline and noradrenaline) and also some of the amines which occur in food. Some of these food amines act to release large amounts of adrenaline and noradrenaline. In the absence of the enzymes to destroy the amines, a substance such as tyramine which occurs in a number of foods can produce a dramatic attack of high blood pressure, with palpitations and a throbbing headache. Patients on these drugs must therefore be carefully instructed to avoid certain foods, including cheese, bananas and hydrolysed meat extracts.

Food allergies

Some unfortunate people are highly sensitive to substances such as eggs or sea foods. If they accidentally eat the forbidden material they may feel weak and ill and come out in an unpleasant rash. The cause of this situation is by no means fully understood. However, it is known that the body can recognize proteins which are foreign to it. It can mount what is called an immune response against the foreign protein and destroy it and this behaviour is of great value in the resistance to bacterial infections. Unfortunately, as a side effect of the immune response, many tissues in the body may become damaged and the patient may become ill.

Normally, of course, protein is completely broken down in the gut. Protein as such cannot be absorbed and gain access to the body: it is first broken down to amino acids which cannot provoke an immune response. Some unfortunate individuals seem to have unusually permeable gut walls and so some proteins may be able to pass through without first being broken down. These proteins can then provoke an immune response. The patient therefore develops an allergy to such proteins and must strictly avoid eating any food which contains them.

Metabolic errors

Some patients have genetically transmitted inborn errors of metabolism. That is to say they lack certain enzymes essential for particular biochemical reactions. A number of these errors present dietary problems. The main ones of importance are:

1. COELIAC DISEASE. Wheat flour contains a protein known as gluten. In some individuals this cannot be fully digested and one of the products of its partial digestion is highly toxic to the wall of the intestine. The gut wall is so damaged that it ceases to be able to absorb a whole variety of materials. Much of the food passes straight through, causing diarrhoea and malnutrition. The disease can be

F

completely cured by a diet which contains no wheat or wheat products and the gut wall then quickly returns to normal.

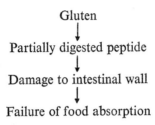

Fig. 9.3. The sequence of events in coeliac disease.

2. DISACCHARIDASE DEFICIENCY. There are a number of enzymes
in the gut which are essential for the breaking down to monosaccharides of the disaccharides sucrose, lactose and maltose. A
deficiency of any one of these enzymes can lead to diarrhoea and a
failure of absorption. If it is lactose which cannot be digested, the
condition appears almost immediately after birth when the infant
takes milk. It can be cured by giving a lactose-free diet, which of
course means avoiding milk. Sucrose enzyme deficiencies tend to
become apparent later when sucrose is first given: they can be cured
by eliminating sucrose from the diet. In these cases, enzyme systems
often develop as the child grows older and so he grows out of his
disease.

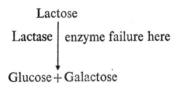

Fig. 9.4. The enzyme failure in lactase deficiency.

3. GALACTOSAEMIA. Some infants cannot convert galactose to
glucose. This is very important because half the carbohydrate taken
in is in the form of galactose from the lactose of milk. The galactose
accumulates and can cause severe brain damage unless it is excluded
from the diet.

4. PHENYLKETONURIA. The body is unable to deal with phenyl-

alanine normally. The amino acid accumulates and abnormal substances derived from it damage the brain. The situation can be improved by giving a diet low in phenylalanine. This is very difficult to do and very expensive because phenylalanine is found in most proteins.

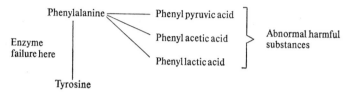

Fig. 9.5. The enzyme failure in phenylketonuria.

Intravenous feeding

Many patients in hospital cannot digest and absorb food normally for a variety of reasons and so they must be maintained by dripping fluid directly into the blood. The important points to remember are:

1. Adequate amounts of fluid must be given. If the patient's kidneys are functioning normally it does not matter if rather too much water is given as the excess can easily be eliminated. But if the patient's kidneys are failing then it is vital to calculate the fluid requirement very carefully as it is easy to give too much water and overload the circulatory system.

2. Sufficient calories must be provided. This is not easy to do, especially if the patient is in renal failure, so restricting the amount of fluid which can be dripped into the body. Highly concentrated glucose solution can be employed, but this often clots the blood in the vein unless the drip is so arranged that it enters a major vessel like the superior vena cava where it can be rapidly diluted. In recent years liquid-fat drips have become available and these can provide a high calorie content in a small volume of liquid.

3. Protein can be provided in the form of predigested amino acids.

4. Elements. Particular care must be taken of the balance of intake and output of sodium, potassium, calcium and magnesium. It is all too easy to provide too little or too much of these elements.

5. Vitamins must be provided by intramuscular injection.

10

The thyroid gland

The hormonal or endocrine system represents one of the two great control systems in the body. The nervous system is the other. The nervous system exerts its control by actually sending nerve fibres to the various organs, but the hormonal system exerts its control at a distance. The endocrine glands release their secretions (hormones or 'chemical messengers') directly into the blood stream.

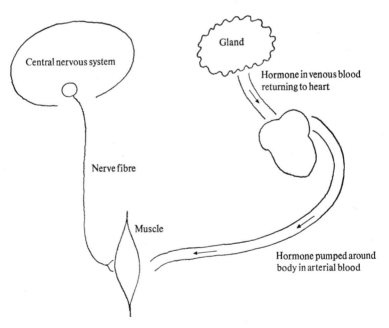

Fig. 10.1. The differences between control of the behaviour of a muscle fibre by the nervous system and by the endocrine system.

The blood then carries the hormones to every part of the body. The nervous and endocrine systems, although apparently so different, are not entirely separate. They come together in the connection between the hypothalamus and the pituitary gland (see chapter 14). The thyroid is perhaps the best known of all the endocrine glands, partly because of its position at the front of the neck and partly because disorders of it are relatively common. It consists of two lobes lying on either side of the trachea (windpipe) and joined by a narrow strip called the isthmus. It receives a very copious blood supply. Surgically it is important to remember that the four parathyroid glands lie embedded on its posterior surface and that the recurrent laryngeal nerve which supplies many of the muscles of the larynx lies very close. Both parathyroids and the nerve may be easily damaged at operation.

Iodine and thyroid hormone

Iodine is essential for the manufacture of thyroid hormone. It is found primarily in sea food and in the sea, and the quantities in food and drinking water diminish on moving inland. It is well absorbed from the gut and as it circulates in the blood it is trapped

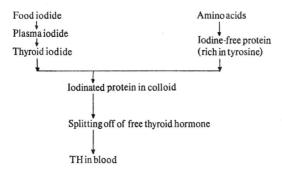

Fig. 10.2. The manufacture of thyroid hormone.

and concentrated in the cells of the thyroid gland. There it is attached to a protein containing the amino acid tyrosine. This iodinated protein is stored in a material called colloid. The colloid itself is found in the centre of spheres of thyroid cells known as follicles. When the hormone is required, it is split off from the colloid and released into the blood. Therefore in an actively secreting gland there is little colloid and the follicles are small. In a resting gland the colloid accumulates and the follicles become large.

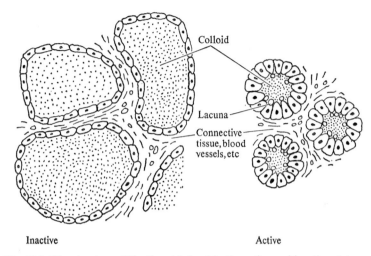

Fig. 10.3. The structure of the thyroid gland in the active and inactive states.

The thyroid hormone is secreted in two forms, thyroxine which contains four iodine atoms and tri-iodothyronine which contains three: both types contain two tyrosine molecules attached together. Over 90 per cent of the hormone is in the form of thyroxine and almost all this is carried in the blood closely bound to the plasma proteins. Therefore if the amount of iodine which is bound to protein is estimated (protein-bound iodine or PBI), this gives a good indication of the level of hormone in the blood. The normal range for the PBI is 3–8 μg/100 ml.

Actions of thyroid hormone

Very many different actions have been described but the most important ones are:

1. Thyroid hormone (TH) can alter the basal metabolic rate. An excess of TH produces a hot, hyperexcitable individual while a lack of TH produces a cold, dull one.

2. TH is essential for normal growth and in this it co-operates with

growth hormone and insulin. However, while growth hormone deficiency leads to a person of small size with normal intellectual and emotional development, TH deficiency leads to severe retardation of growth and of both emotional and mental development.

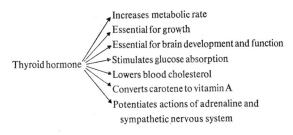

Fig. 10.4. The actions of thyroid hormone.

Children who have suffered from thyroid deficiency since infancy are known as cretins.

3. Even in adults normal blood levels of TH are required for the normal functioning of the nervous system. If TH is deficient the person becomes mentally sluggish but is subject to sudden rages. Excess TH leads to overexcitability and nervousness.

4. TH stimulates glucose absorption from the gut so that very high blood levels of glucose may occur after a carbohydrate meal. This glucose may appear in the urine but it does not indicate diabetes as the blood level usually comes down rapidly again.

5. TH lowers plasma cholesterol levels. If TH is deficient, plasma cholesterol levels may become very high and this is a useful test of thyroid deficiency.

6. TH is required for the conversion of carotene to vitamin A (chapter 8). In the absence of TH carotene accumulates and may give a yellowish tinge to the skin. In severe cases signs of vitamin deficiency may appear.

7. TH increases the effectiveness of adrenaline and of the sympathetic nervous system. These actions are particularly evident in the case of the heart which may beat very rapidly in thyrotoxicosis: it may even start fibrillating.

Control of the thyroid gland

The body clearly requires a method of controlling the amount of TH secreted by the thyroid. It does this with the aid of the pituitary, a tiny gland lying beneath the hypothalamus at the base of the brain.

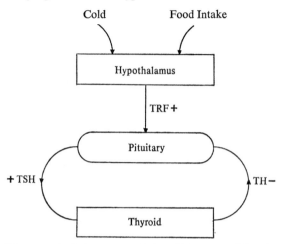

Fig. 10.5. The control of the output of thyroid hormone. Plus signs indicate stimulation, minus signs indicate suppression.

The pituitary exerts an influence on the body out of all proportion to its size. The anterior part of the pituitary releases into the blood a substance known either as thyrotrophic hormone, thyroid stimulating hormone or more usually simply as TSH. An increase in the amount of TSH in the blood stimulates the thyroid to start secreting more hormone. A decrease in the amount of TSH leads to a decrease in the amount of TH released by the thyroid. In turn, the pituitary seems able to measure the amount of TH in the blood and controls its output of TSH accordingly. If the TH level rises, the pituitary releases less TSH so bringing the TH level back to normal. A fall in TH on the other hand stimulates the output of more TSH.

But even the pituitary is not completely independent. It receives blood via a special system of vessels from the hypothalamus. The hypothalamus can secrete into this blood another substance known as TSH-releasing factor or TRF. When the output of TRF rises, so do the outputs of TSH and TH. The secretion of TRF is increased under two main circumstances:

1. Prolonged exposure to cold. The effect is to raise the metabolic rate and so increase the amount of heat produced by the body.

2. Prolonged increase in food intake. Again the metabolic rate tends to increase in order to burn the food up more effectively. Unfortunately this mechanism seems to work much better in some people than in others. Some can therefore eat without getting fat because of their high metabolic rate, while others with a low metabolic rate must watch their weight very carefully.

Clinical aspects

Disorders of the thyroid gland can have four main origins. There may be a lack of iodine in the diet or the hypothalamus, anterior pituitary or thyroid itself may be functioning abnormally. These defects may show themselves in three main ways:

1. Signs of excessive output of thyroid hormone (hyperthyroidism or thyrotoxicosis).

2. Signs of insufficient TH output (hypothyroidism, cretinism in childhood, myxoedema in adults).

3. A swelling in the neck (goitre) without obvious signs of hypo- or hyperthyroidism.

Goitre

This is the term used for a visible swelling of the thyroid. If it is associated with an excessive output of TH it is said to be a toxic goitre. If the output of TH is normal or low, the goitre is said to be non-toxic.

Most goitres are non-toxic and are due to iodine deficiency in the diet. Levels of TH may not be quite sufficient and in an effort to raise them to normal, the pituitary pours out TSH. As a result, the gland is enlarged and if the deficiency is prolonged it may reach an enormous size. In some individuals the output of TH may normally be just sufficient but under the impact of some additional strain such as pregnancy or puberty when additional amounts of TH are required, the gland may again enlarge.

Other goitres may be caused by tumours of the thyroid gland or by an abnormal pituitary which pours out large amounts of TSH even though blood TH levels may be above normal. Such goitres are often toxic.

Hyperthyroidism

This may be caused by overactivity of the hypothalamus, pituitary or thyroid itself and in most cases the precise reason for the disturbance cannot be identified. In yet another group of patients the blood contains a substance of unknown origin called long-acting thyroid stimulator (LATS): this has a similar action to TSH in that it stimulates the thyroid to pour out TH. The main features of hyperthyroidism are:

1. The patient has a high metabolic rate, sweats a lot and dislikes hot weather.

2. The patient is excessively active, both physically and mentally.

3. There is a rapid pulse rate, possibly with atrial fibrillation.

4. The eyeballs protrude. This is partly due to excessive sympathetic nerve activity which pulls on the muscles of the upper eyelid exposing more of the eyeball than usual: this applies in all types of hyperthyroidism. However, in some cases the protrusion is much worse because of the deposition of fat in the orbit behind the eye. In these cases the disease is usually of hypothalamic or anterior pituitary origin and the fat deposition may occur because of the presence of some abnormal anterior pituitary secretion.

There are several different ways of treating thyrotoxicosis. There are a number of antithyroid drugs available (such as carbimazole) which interfere with the manufacture of TH and so reduce its level in the body. Unfortunately the patient usually has to take the drugs for life. Secondly, radioactive iodine can be used. Quite a large dose is employed and the radioactive material is concentrated in the thyroid gland where the radioactivity destroys some of the cells. This is a simple and usually permanent treatment. However, it is difficult to estimate the correct dose of radioactivity and a disturbingly high proportion of the patients treated become myxoedematous later on. The third method of treatment is partial thyroidectomy in which a large part of the gland is surgically removed. Before operation the thyrotoxicosis is brought under control by the use of antithyroid drugs. The blood supply to the gland is also reduced by giving very large doses of iodine (much larger than those normally required in the diet). How the iodine acts is a mystery but it greatly reduces bleeding and makes the operation much easier.

Hypothyroidism

There are two main forms, cretinism in children and myxoedema in adults. Cretinism may be due either to maternal iodine deficiency or to a congenital defect in the thyroid gland which renders it incapable of manufacturing TH. The main characteristics of cretinism are small stature, mental deficiency, an infantile face (because of a failure of bone development), a coarse skin, a slow pulse and sluggish gut. The only treatment is thyroid hormone and a full cure is achieved only if it is started early in infancy.

In adults hypothyroidism is usually termed myxoedema although this refers to only one of its features, the thickening and puffiness of the skin. The skin in these patients in addition to being puffy is dry, waxy and cool. Oedema of the vocal cords leads to a striking deepening and huskiness of the voice. The patient is sluggish both physically and mentally and may be given to sudden rages. The pulse rate is slow and the cholesterol level in the blood is high, a useful diagnostic test. Because of the high cholesterol level the patients seem to be unusually susceptible to fatty changes in the arteries and coronary thrombosis.

11

The adrenal cortex

The adrenal gland consists of two separate parts of quite different functions. The inner medulla is really part of the sympathetic nervous system and is discussed in the physiology book in this series. The outer section of the gland, the adrenal cortex, is very

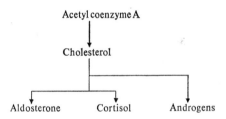

Fig. 11.1. The three types of adrenal cortical hormone are all manufactured from acetyl coenzyme A and cholesterol.

much part of the endocrine system. It produces three different groups of steroid hormones, the mineralocorticoids, glucocorticoids and the androgens.

Mineralocorticoids

The most important of the mineralocorticoids is known as aldosterone. The hormones are so-called because they are primarily concerned with the mineral content of the body, in other words with the inorganic materials. Their main concern is with the body content of sodium ions. They stimulate the kidney to remove sodium from the urine and so to retain it in the body. If there is a large amount of sodium in the body, the output of aldosterone is low, so allowing the excess sodium to escape in the urine. If there is too little sodium in the body, the output of aldosterone is high in order to prevent any unnecessary loss in the urine.

The control of the output of aldosterone is as yet poorly understood. It seems to be little affected by ACTH which is important in

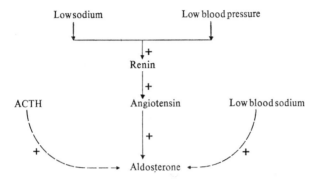

Fig. 11.2. The ways in which aldosterone output may be controlled.

controlling the other adrenal hormones (see later in this chapter). A low sodium concentration in the blood going to the adrenal can raise aldosterone output while a high sodium can reduce the secretion of the hormone: this mechanism makes obvious sense but the changes in concentration required to cause the effect seem large in comparison with those which occur naturally. At present, the most important mechanism of controlling aldosterone output seems to depend on renin, an enzyme which is produced by the kidneys. The output of renin from the kidneys rises if the total sodium content of the body is low: it also rises if the blood pressure is low, as a low blood pressure may be an indication of sodium deficiency. The renin then acts on a plasma protein and breaks off from this protein a substance called angiotensin which is a powerful constrictor of blood vessels. In addition, angiotensin has an action on the adrenal cortex where it stimulates the output of aldosterone. This helps to retain sodium in the body and returns the body sodium content and the blood pressure to normal. In contrast, rises in body sodium content and in arterial pressure can suppress the output of aldosterone by the same renin–angiotensin mechanism, the output of renin being greatly reduced.

Glucocorticoids

Glucocorticoids are so called because they have important actions on carbohydrate metabolism, although that is not their only function. The most important member of the group is cortisol, often known as hydrocortisone. Cortisone is not secreted naturally but is converted to cortisol in the body. Recently some synthetic steroids have been made (e.g. prednisone and prednisolone) which are much more powerful glucocorticoids than cortisol itself. The main actions of the glucocorticoids are:

1. They are essential for the normal excretion of water by the kidney. In their absence the body can get rid of water only slowly.

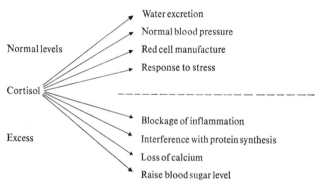

Fig. 11.3. The actions of glucocorticoids such as cortisol and the steroids which are most often used in clinical medicine.

2. They are essential for the maintenance of a normal blood pressure although how they act is unknown.

3. They are required for the manufacture of red blood cells.

4. They are essential to enable the body to respond to any type of stress whether it be pregnancy, a surgical operation, an uncomfortable climate or a bad emotional experience. In all these circumstances the plasma cortisol levels are raised although, again, its precise mode of action is unknown.

In higher concentrations than are normally found in the body the glucocorticoids have other actions as well:

1. They block the inflammatory response. The inflammatory response consists of an increased blood flow and migration of white

blood cells to a damaged area of the body, coupled with a laying down of fibrous tissue. In most cases, such as in the healing of wounds or in fighting bacterial infections, this response is beneficial to the body. But in some cases such as the inflammation of the joints in rheumatoid arthritis or the inflammation of the heart valves in rheumatic fever, the process is harmful: it serves no useful purpose as there are not usually any bacteria in the joints or on the heart valves. Steroids such as cortisol may therefore be used to suppress this harmful type of inflammation. Inevitably, at the same time, they reduce the body's ability to cope with infections, especially tuberculosis. Those who are treated with steroids for long periods of time should therefore have regular chest X-rays.

2. They interfere with the manufacture of proteins and so muscles become weak. The protein collagen, so essential for the strength of blood vessels and of bone is also weak. Bleeding therefore tends to occur and skeletal defects may appear.

3. They cause loss of calcium and phosphate from the kidney. The blood level of calcium is maintained by removal of calcium from bone and this further aggravates the weakness of bone. This weakness is particularly apparent in the vertebral column and in severe cases one or more vertebrae may collapse.

4. They raise the blood glucose level and cause large amounts of fat to be deposited, especially around the shoulders and over the abdomen.

Control of glucocorticoid output

The control of cortisol secretion is similar in many ways to the control of TH secretion. The anterior pituitary secretes a substance known as adrenal corticotrophic hormone or ACTH. ACTH stimulates the adrenal to secrete cortisol. In turn, the ACTH output depends on the secretion of ACTH releasing factor (usually called CRF) from the hypothalamus. Finally the output of CRF depends on the blood level of cortisol. If plasma cortisol levels are too high, the output of CRF falls and with it the outputs of ACTH and cortisol, so returning the plasma cortisol level to normal. If cortisol levels are too low, the output of CRF and ACTH rises, so bringing cortisol levels up again. During periods of stress, the brain directly increases the output of CRF so making more cortisol available to cope with the emergency.

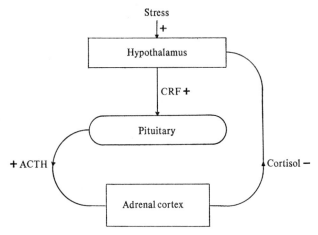

Fig. 11.4. The control of cortisol output.

Androgens

The word androgen means a substance which has an action like that of the male sex hormones. However, despite this name, the adrenal androgens seem to be important in both sexes although relatively little is known about their function. Their main actions appear to be:

1. Stimulation of pubic and axillary hair growth at puberty. Hair grows at puberty in females even in the absence of ovaries (Turner's syndrome) and in males even in the absence of testes.

2. The stimulation of grease production by the skin. Some think that acne is caused by excessive production of adrenal androgens.

3. The development of muscles.

4. The development of sexual desire in both males and females.

CLINICAL ASPECTS OF ADRENAL FUNCTION

A number of clinical conditions are associated with disorders of the adrenal glands. The main ones are discussed in this section.

Addison's disease

This is due to a failure of the adrenal to secrete its hormones. It is usually due to destruction of the gland, often by tuberculosis but frequently for unknown reasons. The most important consequences are a massive loss of sodium in the urine coupled with a very low blood pressure. In chronic cases, anaemia, low blood glucose, failure to respond to stress and pigmentation of the skin are also apparent. The skin pigmentation develops because the pituitary pours out ACTH in an effort to raise the output of adrenal hormones to normal. ACTH has a skin-darkening action in large doses and results in a peculiar bronzed appearance, giving a false impression of good health. The disease can be fairly satisfactorily treated by means of the various hormone preparations available.

Conn's syndrome

This is due to a tumour which appears to pour out excessive amounts of mineralocorticoids alone. As a result there is salt retention and high blood pressure. Because of the way in which the kidney works, the retention of large amounts of salt in the body is often associated with a loss of potassium in the urine. Thus the blood level of potassium is often low resulting in cramps and muscular weakness. The condition can be cured by removal of the tumour.

Cushing's syndrome

This is primarily due to an excess secretion of glucocorticoids although mineralocorticoids and androgens are often also secreted in above normal amounts. It may occur because of an adrenal tumour, because of an excess output of ACTH from the pituitary, or because of a cancer which for no apparent reason suddenly starts pouring out large amounts of ACTH. Lung cancers are the commonest type to behave in this way (chapter 6). The main features of Cushing's syndrome are:

1. A high blood glucose, coupled with obesity. The limbs tend to be relatively thin, contrasting oddly with the grossly fat face and trunk.

2. Suppression of the inflammatory response with a resulting susceptibility to infections.

3. Easy bruising and bleeding with weak muscles.

4. Backache because of loss of calcium from the vertebral column.

5. High blood pressure.

6. A high red cell count.

7. The development of hair on the face and other parts of the body. This is naturally more obvious in females.

G

Androgenic excess

This may occur in early infancy or even in utero when it is usually due to a congenital defect in one of the enzymes manufacturing aldosterone or cortisol. As a result these hormones cannot be manufactured and the adrenal cortex devotes all its energies to making androgens. In female children, the external genitalia become masculinized with growth of the clitoris. Male genitalia hypertrophy. In both sexes hair development and muscle growth occur as in adults. In later childhood, similar changes may be brought about by androgen-secreting tumours. In adult males androgen-secreting tumours often go unnoticed until a late stage but in females they cause pronounced masculinization early on.

12

Calcium and the parathyroids

Calcium is important in the body in many different ways. The main ones are:

1. It is essential for the normal activity of many enzymes.

2. It is important in maintaining the stability of cell membranes.

3. It is required for muscular contraction and for the normal functioning of the nervous system.

4. It is essential for blood clotting.

5. It is essential for the manufacture of strong bones and teeth.

The total concentration of calcium in the plasma is 9–11 mg/100 ml. It exists in two main forms, as free calcium ions and as calcium bound to plasma protein. The two types can interact as follows:

$$\text{Calcium ions} + \text{protein} \rightleftharpoons \text{Calcium-protein complex}$$

Only the free calcium ions are physiologically active. The balance between free and protein-pound calcium depends largely on the pH of the blood. If for any reason the blood becomes more alkaline, more calcium is bound by the protein and the concentration of free calcium ions may fall below normal. The main situations when the blood may become more alkaline are:

1. When acid is lost because of repeated vomiting of gastric juice.

2. When excessive amounts of the potentially acid gas, carbon dioxide, are lost from the body because of overbreathing. This is most likely to occur in hysteria and during childbirth when the mother often overbreathes.

The concentration of free calcium ions will also fall if the total plasma calcium falls. This occurs when the parathyroid glands fail to function normally. Whenever the concentration of free calcium ions falls, whatever its cause, the condition of tetany results. The symptoms are due to overactivity of the nervous system. Nerve impulses fire off spontaneously in both motor and sensory nerves. The abnormal sensory impulses cause tingling sensations often known as 'paraesthesiae' while the motor impulses cause involuntary muscle twitches. These twitches are usually most apparent in the muscles of the inner side of the hand and forearm supplied by the ulnar nerve. As a result the fourth and fifth fingers curl up giving the 'main d'accoucheur' so called because it is the position of the hand when doing a vaginal examination. The facial muscles also often twitch.

When due to overbreathing, tetany is very easily cured by simply putting a paper bag over the mouth and nose for a few minutes. The carbon dioxide breathed out is then breathed straight back in again. This makes the blood less alkaline and the link between calcium and protein is broken. The concentration of free calcium ions returns to its normal level and the tetany disappears. Tetany due to a low calcium concentration may be rapidly but temporarily relieved by injecting a solution of a calcium salt (e.g. calcium gluconate) intravenously.

Bone

Bone contains three main types of cell and three main types of extra-cellular material. The basic structure of bone consists of a thick interlacing web of fibres of the protein collagen. These fibres are glued together by a special type of carbohydrate known as a muco-polysaccharide. Finally they are made hard and durable by the deposition on their surfaces of mineral substances, primarily calcium and phosphate. The three main types of cell are the osteocytes, osteoblasts and osteoclasts. The osteocytes predominate in bone which is stable and is neither being built up nor destroyed: they seem in an unknown way to maintain the health of the bone. The osteo-blasts predominate where new bone is being laid down while the osteoclasts are large multinucleated cells which seem to be involved in the breakdown of old bone.

There are two main types of bone, the long bones and the flat bones. The flat bones grow because their inner surfaces are eroded while new bone is laid down on their outer surface. The long bones grow primarily at the epiphysial regions. Each long bone has at least at one end, and often at both, a separate piece of bone known as the epiphysis. New bone-formation occurs mainly in the region between the shaft of the long bone and its epiphysis. This region is rich in cartilage rather than bone. After puberty when growth ceases, the epiphyses become firmly united to the shafts by proper bone.

Calcium absorption

The main sources of calcium are dairy foods. Like iron, calcium is poorly absorbed but there are two factors which help the process. Vitamin D is the most important of these. A normal secretion of acid gastric juice is helpful but not essential: this is because in an

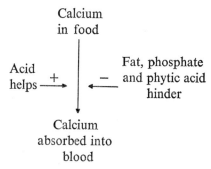

Fig. 12.1. Factors affecting the absorption of calcium from the gut.

alkaline medium calcium–phosphate combinations tend to be very insoluble while they are more soluble as the pH falls. Thus patients who have little natural acid secretion or whose stomachs have been removed at operation are in danger of becoming calcium deficient.

Three main factors can hinder the normal absorption of calcium:

1. The presence of large amounts of unabsorbable fat in the gut. The fat forms an insoluble complex with the calcium and stops the calcium being absorbed as well. This situation may occur when fat is poorly digested because of lack of bile or of pancreatic juice. It may also occur if the wall of the intestine is damaged as in sprue or coeliac disease.

2. The presence of large amounts of phosphate in the food which also forms an insoluble combination with calcium.

3. The presence of excess phytic acid in food which yet again forms an insoluble complex. Phytic acid is found primarily in wheat flour and in order to combat its action, extra calcium is added to the bread in many countries.

Vitamin D

This has already been discussed in chapter 8. It has two main actions on calcium metabolism. In the gut it is essential for calcium absorption and in its absence enough calcium cannot be taken into the body. Once in the body, it is required for the normal manufacture and calcification of new growing bone, especially at the epiphyses. In its absence normal bone growth cannot occur, giving the disease of rickets. The long bones and the pelvis in particular may become permanently deformed.

Parathyroid hormone

The parathyroid glands are essential for the maintenance of normal plasma calcium levels. They are four tiny structures normally embedded in the back of the thyroid. Not unusually, however, one or more may be found elsewhere in the neck or in the thorax. The glands secrete a hormone known as parathormone which has three main actions:

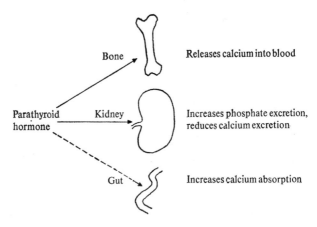

Fig. 12.2. The actions of parathyroid hormone.

1. It releases calcium from bone into the blood.

2. It reduces the excretion of calcium by the kidneys. These first and second actions tend to raise the plasma calcium level.

3. It causes the excretion of phosphate by the kidneys. This lowers the plasma phosphate level and is important because calcium and phosphate together tend to precipitate out of solution as an insoluble complex. This can damage many tissues and arterial walls in particular. This is obviously a highly undesirable event and it can be prevented if the phosphate concentration is lowered as the calcium concentration in the plasma rises. This is the action of parathormone.

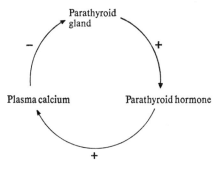

Fig. 12.3. The control of blood calcium levels.

Parathormone may have a minor action in increasing calcium absorption from the gut but this is not thought to be very important. The output of parathormone depends on the level of calcium in the blood. If the plasma calcium falls, the output of parathormone increases and raises the calcium back to normal. If the plasma calcium rises, the secretion of parathormone is suppressed so allowing calcium levels to fall back to normal.

Calcitonin

This is a relatively recently discovered hormone which may be made by both thyroid and parathyroid glands. Its action is to lower plasma calcium by causing the deposition of calcium in bone. Its importance in medicine has not yet been worked out.

Parathyroid Disorders

Underactivity of the parathyroids (hypoparathyroidism) may be caused by atrophy of the glands for unknown reasons or by accidental removal of the glands during partial thyroidectomy. The plasma calcium level falls and tetany results. The tetany may be temporarily relieved by an intravenous injection of calcium gluconate. In the long term the condition must be treated by a very high calcium diet and large doses of vitamin D.

Hyperparathyroidism is much rarer and is usually due to a tumour. It is commoner in women. Calcium is removed from the bones causing weakness: it is then deposited elsewhere in the body, particularly in arterial walls, in the kidneys and in the urine. For unknown reasons peptic ulcers are particularly common. The only treatment is removal of the tumour.

13

Regulation of blood glucose

Normally the concentration of blood glucose is within the range 60–90 mg/100 ml of plasma. The maintenance of a reasonably constant level is one of the most important functions of the hormonal system. This is so that at all times all organs of the body can obtain adequate supplies of food from which they can obtain energy. The constancy is particularly important for the cells of the brain which seem able to oxidize glucose only and which cannot make use of fat. Almost all the cells of the body apart from those of the liver and brain present some barrier to the entry of glucose from the blood. Many of the hormones act by altering the effectiveness of this barrier so changing the ease with which the glucose can enter the cells. Cells of the liver and brain are freely permeable to glucose at all times.

The main problem of blood glucose regulation results from the fact that meals are taken relatively infrequently. Immediately after a meal large amounts of glucose enter the blood and the plasma concentration tends to rise. But within a short time the absorption of the meal is completed and the blood glucose level tends to become very low. The aim of the regulating system is to iron out these wild swings, to lower the blood glucose level just after a meal and to raise it between meals.

The first line of defence against these swings is provided by the liver which receives all the portal vein blood from the gut. When glucose is being absorbed very rapidly just after a meal, much of it is taken up by the liver and stored in the form of glycogen: the blood leaving the liver via the hepatic veins contains less glucose than the blood entering the liver via the portal vein. In contrast, when

absorption has been completed and no glucose is entering the portal
blood from the gut, the liver breaks down some of its glycogen and
releases it into the blood as glucose. In this way a sharp fall in blood
glucose can be avoided.

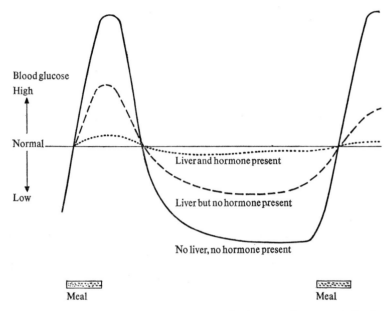

Fig. 13.1. The blood glucose level is normally kept nearly constant. In the
absence of the liver and of hormonal control the wild swings shown in the
diagram would occur.

The liver is aided in this smoothing-out action by the activities of
a number of hormones.

Insulin

This is a protein hormone released by tiny groups of cells in the
pancreas known as the islets of Langerhans. The islets are quite
distinct in structure from the main part of the pancreas which
secretes digestive juices along the pancreatic duct. The islets secrete
their hormone directly into the blood. The most important action
of insulin is to lower the blood sugar. It does this in two ways:

1. It stimulates the liver to manufacture glycogen from the glucose
which reaches the liver in the blood.

2. It stimulates most of the cells in the body, particularly those in muscles, to take up glucose quickly. In the absence of insulin many cells in the body (but not those of the brain and liver) become almost impermeable to glucose and cannot take it up from the blood. They

Fig. 13.2. The actions of insulin.

must therefore use fat instead for their energy supply. In skeletal muscles and in the heart much of the glucose taken into the cells is converted into a glycogen store.

The other main actions of insulin are to stimulate the manufacture of proteins from the amino acids absorbed after a meal and to stimulate the formation of triglycerides in adipose tissue from the fatty acids in the blood. Insulin therefore tends to increase the availability of carbohydrate and to reduce the availability of fat.

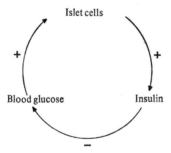

Fig. 13.3. The control of insulin output.

The output of insulin is controlled by the level of blood glucose itself. When the blood glucose level rises, as after a meal, the insulin output also rises, so tending to bring the blood glucose back to normal. When the blood glucose falls, the rate of insulin secretion also falls: this helps to stop the cells using glucose and so the blood glucose level tends to rise again.

Growth hormone

This is secreted by the anterior pituitary gland and in some ways its actions are the reverse of the actions of insulin. The main ones are:

1. It stops most of the cells in the body (like those of muscle) taking glucose from the blood. This halts the fall in blood glucose. It has no effect on the brain cells and so there is plenty of glucose available for the nervous system.

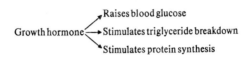

Fig. 13.4. The actions of growth hormone.

2. It causes the breakdown of fat to free fatty acids. These enter the blood and most cells, muscle in particular, use them as an alternative energy supply.

3. It continues to stimulate protein synthesis and so does not interfere with this important aspect of metabolism.

The output of growth hormone is also controlled by the blood glucose level but the effects are precisely opposite to those which control insulin output. The output of growth hormone is increased when the blood glucose level falls and decreased when the blood glucose level rises. So just after a meal when blood glucose concentration tends to be high, insulin levels are also high and growth hormone output is low. The high insulin concentration stimulates the cells to use glucose. But when the meal has been absorbed and blood glucose levels fall, the insulin level also falls while growth hormone output increases. Thus there is a barrier to the entry of

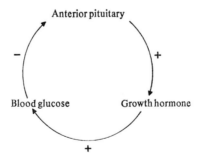

Fig. 13.5. The control of growth hormone output.

glucose into cells but an increase in the availability of fats. The body thus tends to use fat instead of glucose: at the same time the glucose concentration in the blood is maintained for the use of the brain cells which can use nothing else. The day to day control of glucose levels and the balance between the uses of fat and of carbohydrate depend primarily on a close co-operation between insulin and growth hormone.

Adrenaline

This is not routinely important in the control of blood glucose but it can raise the concentration rapidly in an emergency when the brain is in danger of not getting enough energy. It acts in two ways:

1. It stimulates the liver to break down glycogen to glucose quickly.

2. It reduces the rate at which cells like those of muscle remove glucose from the blood.

These actions produce a rapid rise in blood glucose so that cells in the nervous system can receive an adequate supply. The output of adrenaline is stimulated by a low blood glucose concentration and the release is accompanied of course by other actions of adrenaline and of the sympathetic nervous system. A hypoglycaemic (low blood glucose) attack is therefore accompanied by sweating and by a thumping, rapid heart beat. There is usually a feeling of faintness because of the lack of energy supply to the nerve cells.

Other hormones

Two other hormones can affect glucose metabolism but their precise significance under normal conditions is uncertain.

1. Glucagon is also produced by the islets of Langerhans but it has precisely the opposite actions to insulin. It raises blood sugar and it even stimulates the breakdown of protein to give glucose so that it has been suggested that it may be of use in starvation.

2. Cortisol. High levels of cortisol produce a high blood glucose as in Cushing's syndrome. Low cortisol levels as in Addison's disease lead to a low blood glucose level. Despite this the precise role of cortisol in glucose metabolism is uncertain.

Diabetes mellitus

The word diabetes used to be used for any condition in which the output of urine was excessive. Two varieties were known; in one the urine was sweet (mellitus) and in the other it was tasteless (insipidus). Diabetes insipidus is much rarer than diabetes mellitus and is due to lack of antidiuretic hormone from the pituitary (see next chapter). When the word diabetes is used alone it usually refers to diabetes mellitus.

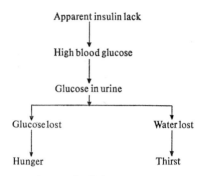

Fig. 13.6. The sequence of events in diabetes.

Diabetes mellitus is a state in which the insulin levels appear to be too low to cope with the patient's intake of carbohydrate. As a result five important things tend to occur:

1. The blood sugar level is much higher than normal.

2. If the blood sugar level rises above 180 mg/100 ml, the kidneys cannot hold glucose back from the urine and so glucose appears in the urine. This is abnormal and the glucose can be picked up by several different tests (Benedict's, Fehling's, 'Clinitest' and 'Clinistix').

3. When glucose appears in the urine, excess water must also be excreted to carry away the glucose in solution. The patient therefore complains about producing large amounts of urine.

4. The large amounts of water lost make the patient very thirsty and he complains of drinking a lot.

5. Because much of the glucose taken into the body is lost, more calories must be supplied by excessive eating.

In early diabetes all these features may not be present and there may be no sugar in the urine if it is tested at random. The diabetic state may then be uncovered by means of a glucose tolerance test.

In this, a patient is given a large amount of glucose either by mouth or intravenously. Urine and blood samples are then taken at half-hourly intervals for 2½–3 hours. Normally the blood glucose concentration does not rise above about 150 mg/100 ml and then rapidly

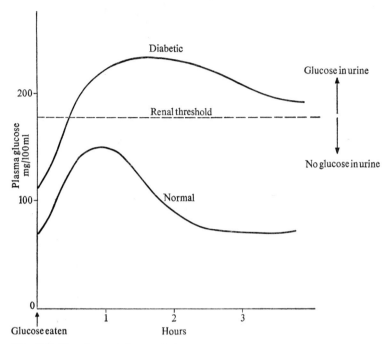

Fig. 13.7. The glucose tolerance test.

returns to normal: no glucose appears in the urine. In diabetes the blood glucose becomes very high, stays high for a long period, and glucose overflows into the urine.

In addition to the features mentioned above, diabetes may cause many complications most of which are not fully understood. The main ones are:

1. Deposition of fat in arteries, leading to heart disease, and the blocking of major arteries, especially in the legs.

2. Damage to the kidneys ending in renal failure.

3. Damage to the retina ending in blindness.

4. Repeated infections, especially of the skin and urine. This seems

to be because bacteria find the glucose-rich body fluids an ideal place to grow.

5. Ketosis and diabetic coma. This is perhaps the best-known complication and its mechanism is now fairly well understood. It begins because there is not enough insulin to enable most of the cells in the body to use carbohydrate: as a result there is a switch to the oxidation of fat instead. Because of the lack of carbohydrate breakdown there may not be enough oxaloacetic acid to combine with acetyl coenzyme A to allow for complete fat breakdown by the citric acid cycle. The accumulation of acetyl coenzyme A leads to the formation of the ketone bodies (chapter 4): these make the blood acid and in addition are poisonous to the brain. In consequence, the patient becomes comatose. There are usually three vital things wrong with a patient in diabetic coma:

    a. The high levels of toxic ketone bodies.

    b. The acidity of the blood.

    c. The high levels of blood glucose which usually cause massive fluid loss in the urine so that the patient is severely dehydrated.

The treatment is therefore directed to correcting these three things.

    a. Insulin is given in large amounts to switch the body metabolism to carbohydrate use and to burn up the ketone bodies.

    b. Intravenous fluids are given to make up the deficiency.

    c. Some of the fluid is usually given in the form of alkaline bicarbonate, lactate or citrate solutions which help to counteract the acidity of the blood.

Types of diabetes

Diabetics fall into two quite distinct categories. The so-called 'juvenile' type genuinely do have a lack of insulin. They are usually very thin because of the large losses of glucose and can be treated only by injections of insulin. Insulin cannot be given by mouth because it is digested in the gut.

In the much commoner 'adult' or 'mature' type, the main problem does not seem to be an insulin lack. In fact insulin levels are often higher than in normal individuals. Such diabetics tend to be obese. There are several possible explanations for this mature type, none of which is fully accepted.

1. There may be a very high carbohydrate intake which exceeds the capacity of insulin to cope with it.

2. The cells may be resistant to the action of insulin so that much higher levels of the hormone may be required to produce a normal effect.

3. There may be present in the blood, insulin antagonists which prevent insulin exerting its normal action. This certainly seems able to account for the type of diabetes seen in Cushing's syndrome when the excess cortisol levels persistently push up the plasma glucose levels. An increased output of growth hormone which also tends to raise blood glucose may explain some other cases. For example, it is well-known that a baby whose birth weight is over 10 pounds almost always indicates that the mother is a diabetic or will develop diabetes sooner or later. This could be due to an excessive maternal output of growth hormone.

Since mature diabetes is not due to an actual absolute lack of insulin it may be possible to treat it without insulin injections. In fact there are four main ways of tackling it:

1. By cutting down the intake of carbohydrate in the food (especially of sucrose) so reducing the need for insulin.

2. By using oral drugs like tolbutamide which stimulate the islets of Langerhans to pour out more insulin.

3. By using oral drugs like phenformin which have an insulin-like action and which stimulate the uptake of glucose by cells.

4. By insulin injections. In practice most mature diabetics can be satisfactorily treated by some combination of the first three measures, so avoiding the need for injections of insulin.

H

14

The pituitary and hypothalamus

The pituitary is the single most important endocrine gland in the whole body. This is because in addition to producing hormones which act in their own right, it also manufactures hormones (so-called trophic hormones) which control the behaviour of the thyroid, the adrenal cortex, the ovaries and the testes. In turn the pituitary itself is controlled by the part of the brain known as the hypothalamus and so in the end the brain is intimately involved in controlling the behaviour of the endocrine system.

The pituitary has two quite distinct parts, anterior and posterior, sometimes called the adenohypophysis and the neurohypophysis. During foetal life, the posterior pituitary develops as a downgrowth from the hypothalamus. The anterior pituitary develops as an up-

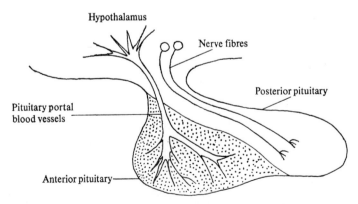

Fig. 14.1. The structure, blood supply and nerve supply of the pituitary gland.

growth from the roof of the mouth. The posterior pituitary receives a very rich nerve supply from the hypothalamus. The anterior pituitary seems to have no nerve supply but instead is intimately connected to the hypothalamus by a complex system of blood vessels known as the pituitary portal system. The hypothalamus releases chemicals known as releasing factors (RF) into these vessels. They are then carried by the blood down the pituitary stalk to the anterior pituitary where they stimulate the output of the anterior pituitary hormones.

THE ANTERIOR PITUITARY

This produces two hormones which act in their own right and four trophic hormones which act by controlling the behaviour of other endocrine glands.

Growth hormone

This is essential for the normal growth of a child. If it is deficient, a dwarf results: if it is present in excess the outcome is a giant. Apart from their size, these individuals are normal human beings. This is in marked contrast to the retardation of mental and emotional development which occurs in thyroid deficiency. If excessive output of growth hormone occurs after the epiphyses of the long bones have fused and growth in height has ceased, growth hormone can then have much effect only on the soft tissues and on the bones of the hands, feet and face. Thus the nose, hands, feet and soft tissues such as the tongue may become enormous in size. This condition is known as acromegaly.

Apart from its action in stimulating growth, growth hormone is also important in the regulation of blood sugar (chapter 13). It is essential for the switch from predominantly carbohydrate to predominantly fat oxidation which occurs once a meal has been completely absorbed. In acromegaly when the output of growth hormone is persistently high, diabetes frequently occurs: the growth hormone keeps the blood glucose level high and in an attempt to secrete enough insulin to keep the glucose level down, the islets of Langerhans appear to become exhausted.

Prolactin

This is essential for the development of the breasts during pregnancy

H*

and for the secretion of milk once birth has taken place. The output of prolactin is stimulated by suckling. The sensory nerves in the nipples send nerve impulses up to the hypothalamus which then

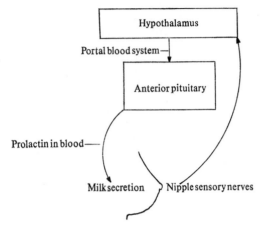

Fig. 14.2. The stimulation of milk secretion by prolactin.

stimulates the output of prolactin from the anterior pituitary. If suckling stops, prolactin secretion falls to very low levels and within a short time the breasts cease to secrete milk.

Adrenal corticotrophic hormone (ACTH)

This is essential for the normal control of the output of cortisol from the adrenal cortex. It may also help in the stimulation of the output of aldosterone and the androgens but in these cases it probably does not act alone. In turn, the output of ACTH is controlled by the secretion of corticotrophin releasing factor (CRF) by the hypothalamus into the pituitary portal blood vessels.

Thyroid stimulating hormone (TSH)

This stimulates the thyroid gland to secrete thyroid hormone. The output of TSH itself is in turn controlled by the level of thyroid hormone in the blood and also by the level of the TSH releasing factor (TRF) secreted by the hypothalamus (chapter 10).

Follicle stimulating hormone (FSH)

This is the hormone which is essential for the development of the egg-bearing follicles in the ovary. It is also required for the manufacture of sperm by the testis in the male. Even though it is called FSH after its action in the female, it is now known that the FSH in males is identical to that in females: it is the ovaries and testes which differ in their response to it.

Luteinizing hormone (LH)

This is the hormone which stimulates the output of hormones from the ovary and from the testis. The ovary secretes primarily oestrogens and progesterone while the testis secretes testosterone. Again LH is identical in males and females: in males it is sometimes called interstitial cell-stimulating hormone or ICSH. The actions of these sex hormones are further discussed in the text on physiology.

Sometimes, since prolactin, LH and FSH are all concerned with sexual function, they are lumped together under the name of gonadotrophic hormones.

THE POSTERIOR PITUITARY

This produces two important hormones, antidiuretic hormone (ADH or vasopressin) and oxytocin. Both these hormones are made by nerve cells in the hypothalamus. They then travel down the nerve fibres which pass along the pituitary stalk to the posterior pituitary. The hormones are actually released into the blood in the posterior pituitary gland.

Oxytocin

This is important in three ways. During delivery of a baby, sensory receptors in the wall of the uterus and in particular in the cervix are activated. They send nerve impulses up to the hypothalamus which then stimulates the output of oxytocin from the posterior pituitary. The oxytocin makes the uterus contract more vigorously, so helping to push the baby out.

Secondly, oxytocin is essential for the ejection of milk from the breasts. Prolactin stimulates the manufacture of milk but it cannot expel the milk from the breasts out of the nipples. The milk simply accumulates in the ducts in the glands and makes the breasts en-

gorged and painful. However, the ducts are surrounded by cells
with a muscular type of action known as myoepithelial cells. When
the baby sucks the nipple, sensory receptors in the nipple are
activated. They send nerve impulses to the hypothalamus which

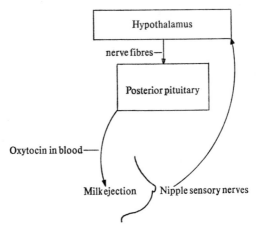

Fig. 14.3. The stimulation of milk ejection by oxytocin.

then stimulates oxytocin secretion. The oxytocin travels in the blood
to the breasts and makes the myoepithelial cells contract, thus
forcing milk out of the breast. This is why when a baby is suckling
or a cow is being milked, it takes a minute or two before the milk
flows freely.

Lastly, oxytocin appears to be important during sexual intercourse.
It is probably partly responsible for the female sensation of orgasm
which is associated with contractions of the uterus.

Oxytocin is also present in males but its function, if any, is
uncertain.

Antidiuretic hormone

This hormone acts on the kidney to reduce the output of water in
the urine. If the body water content is low, the plasma becomes
slightly concentrated. This change in plasma concentration is
detected by the hypothalamus which increases the output of ADH.
The ADH travels in the blood to the kidney and reduces the amount
of water lost in the urine, making the urine smaller in volume and

more concentrated. On the other hand if an excess of fluid is drunk, this dilutes the blood a little. This change too is detected by the hypothalamus which reduces the amount of ADH secreted, so allowing the kidney to excrete a large volume of dilute urine. If

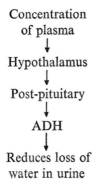

Concentration
of plasma
↓
Hypothalamus
↓
Post-pituitary
↓
ADH
↓
Reduces loss of
water in urine

Fig. 14.4. The control of ADH output.

ADH is absent because of damage to the hypothalamus or posterior pituitary, vast amounts of dilute urine are secreted, a condition known as diabetes insipidus. Because of the tremendous loss of water the patient is perpetually thirsty.

Index

NOTES

NOTES

NOTES

NOTES

NOTES

NOTES